GetSkills

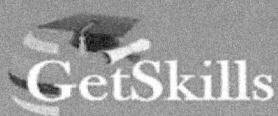

PRECISE SOFTWARE TESTING - ISTQB FOUNDATION STUDY GUIDE

ANZTB
ANZ TESTING BOARD

ISTQB
International Software Testing Qualifications Board

ISTQB
Accredited Training Provider

GETSKILLS Level 27, PWC Tower , 188 Quay Street , Auckland Central, New Zealand

Table of Contents

1. Fundamentals of Testing

1.1 What is Software Testing?

Testing is investigating the project/product with respective to the customer requirements and specifications. It is the process in which defects are detected, isolated and subject for rectification, finally ensuring that the application is error free and working as per the user requirements, specifications in all the conditions.

IEEE defines "Testing is the process of exercising or evaluating a system or a component (module) of a system by manual or Automation means to verify that is satisfies specified user requirements".

Software testing builds the quality in to the application with thorough process of testing Testers will ensure to deliver Error free software to customer. Software testing is also defined as the process Technical investigation provided to stake holders for their application or product which is under test called as either Application under test (AUT) or Software under test (SUT).Basically there are two types of Testing's that are followed by testers in any companies.

Manual Testing: Process in which testing is done manually by executing all the input and output combinations of Application under testing by writing Test cases and scenarios and executing them to find errors.

Automation Testing: When testing is done with the help of tools in any phases to check their functionality or find the errors in it is called Automation testing. Tools like Selenium, SOAP UI, QTP, Meter, Silk Test, and Load Runner etc are used in any stages of testing software.

There are many approaches to software testing there are:

Static testing: Static testing is a software testing method that involves examination of the program's code and its associated documentation but does not require the program be executed. Static testing may be conducted manually or through the use of various software testing tools. Specific types of static software testing include code analysis, inspection, code reviews and walkthroughs.

Dynamic testing: Dynamic testing is a method of assessing the feasibility of a software program by giving input and examining output (I/O). The dynamic method requires that the code be compiled and run.

Software testing is used in association with verification and validation:

Verification: Have we built the software right i.e. Does it match with the customer specifications?

Validation: Have we built the right software i.e. Is this the customer wants?

As per the ISTQB exhaustive testing is impossible means testing all the inputs and pre conditions is not feasible when testing anything other than a simple product. This means that the number of defects in a software product can be very large and defects that occur infrequently are difficult to find in testing. Testing is an artistic job where everyone has their own way and own strategy of testing. So, optimal testing is feasible for testing the applications.

1.2 Why Testing is necessary?

Software Testing is necessary because we all make mistakes. Some of those mistakes are unimportant, but some of them are expensive or dangerous. We need to check everything and anything we produce because things can always go wrong – humans make mistakes all the time.

- Developers are also humans they too can make mistakes due to improper understanding on the requirements. Moreover Developers are not good Testers to test their own work.
- No companies like their customers to find bugs after releasing the product because proper software testing before delivery ensures that software developed is error free and working as per their customer specifications.

- Post release debugging is expensive. Fixing the errors in software in the later stages will cost higher and will have an impact in other modules of the software too.

- Testing is necessary in order to provide the facilities to the customers like the delivery of high quality product or software application which requires lower maintenance cost and hence results into more accurate, consistent and reliable results.

- It might sound dramatic but some of the critical systems without proper testing may result in serious injuries or death. Example: Flight simulation software's, Defense software's.

- If an organization is unable to deliver reliable quality software to their customers it will definitely damage their company's reputation in the market.

1.3 Importance of software testing and major Testing failures occurred in the world:

Software testing plays key role as we discussed in the former sections about the testing importance in gaining the customer confidence and feedback to our company. Let us see the History's great software failures and their impact so, that it gives clear idea how important software testing is in terms of money and good will of a company.

1. November 2000 - National Cancer Institute, Panama City

Multidata Systems International, a U.S. firm developed a therapy planning software miscalculates the dosage given to patients who undergoes radiation therapy. This miscalculations lead eight patients die, while another 20 receive overdoses likely to cause significant health problems. The physicians, who were legally required to double-check the computer's calculations by hand, are indicted for murder.

2. USS Crew 1998

A crew member of the guided-missile cruiser USS Yorktown mistakenly entered a zero for a data value, which resulted in a division by zero). The error cascaded and eventually shut down the ship's propulsion system. The ship was dead in the water for several hours because a program didn't check for valid input. (Reported in Scientific American, November 1998).

3. Ariane 5- 1996

The Ariane 5 rocket exploded on its maiden flight in June 1996 because the navigation package was inherited from the Ariane 4 without proper testing.

4. Heart Bleed Bug -2012.

A security bug in the TCP IP protocol used for internet which injects vulnerabilities in to SSL certificates used for secured browsing made to retrieve sensitive information's of user from the effected websites he is using. As of May 20, 2014, 1.5% of the 800,000 most popular TLS-enabled websites were still vulnerable to Heart bleed. As of May 8, 2014, 318,239 of the public web servers remained vulnerable. According to Net craft, about 30,000 of the 500,000+ X.509 certificates which could have been compromised due to Heart bleed had been reissued by April 11, 2014, although fewer had been revoked.

1.4 Objectives of software testing:

We can use both the static and dynamic testing techniques to achieve our goals and objectives in testing which improves the process of Developing and Testing a project/product. Testing process can have different objectives and goals which may include the three main objectives or goals of software testing are:

1. Fault Identification: What Fault caused the failure?

2. Fault correction: change the system

3. Fault removal: Take out the faults

Apart from the three objectives there are different goals in software testing those are:

- To ensure that application works as specified in requirements document.
- To achieve the customer satisfaction.
- To ensure that error handling has been done gracefully in the application. In situations when user has entered incorrect data, the application display user-friendly error messages.
- To establish confidence in software
- To evaluate properties of software
- To discuss the distinctions between validation testing and defect testing
- To describe strategies for generating system test cases

- To understand the essential characteristics of tool used for test automation

Incase if we are making any changes or enhancements to the software already in use then our testing objective changes and we have to make sure that newly developed features or enhancements should not affect the working of software. We have to ensure that no defects are introduced due to this enhancement by doing regression testing.

If an area of the code is particularly complex and tricky, or because changing software and other products tends to cause knock-on defects. Testers will often use this information when making their risk assessment for planning the tests, and will focus on these tricky components which have more bugs.

Software testing shall ensure that user accepts the final software released for him to operate with no complaints. In order to fulfill this objective, tester shall have end-user mindset which would help in writing test cases or scenarios to meet user expectations.

Software Testing shall ensure that the proper testing is being done to ensure that the system is ready to use. Good testing covers functionality, install ability, and operational ease to learn and use the system. Also, it verifies that the system is easily deployable and replaceable. This would make the system that is easy to install, learn, and use.

Software Testing shall allow to gain confidence that it works. This would happen when system testing proves that the system is reliable and does not crash or there will be no show stoppers.

Software testing shall evaluate the capabilities of a system to show that a system performs as intended. This allows us to understand the limits of performance, land also learning what a system is able to do and not able to do. For the user with good knowledge on the system, the results from the system shall be predictable.

Software testing shall also verify documentation. Many documents are created and also, evolve throughout software development life cycle. Along with this there can be on-line help, installation and troubleshooting related, user training related documents. Software testing shall evaluate all these documents for correctness.

Once the system is operational then our objective of testing is to check the system performance under various loads and its computability, availability and reliability. Reviewing defects and failures in order to improve processes allows us to improve our testing processes.

The main objective of testing is to find the bugs as early as possible in the initial phases of testing which reduces the cost of fixing them and prevents those bugs to reappear in the final stages of deployment.

1.5 Principles of software testing:

There are 7 software testing principles which are universally recognized and accepted. These 7 testing principles are to be followed by the testers in any organization involving in the Testing process. These are the common principles for all types of testing. They are as follows:

1) **Testing shows presence of defects:** Testing can show the defects are present, but cannot prove that there are no defects. Even after testing the application or product thoroughly we cannot say that the product is 100% defect free. Testing always reduces the number of undiscovered defects remaining in the software but even if no defects are found, it is not a proof of correctness. This principle states that Software testing always help to find the defects but it cannot ensure that the software is bug free and error free even after fully testing is done because no one can predict when uncovered or undiscovered bugs will cause system failures.

2) **Exhaustive testing is impossible:** Testing everything including all combinations of inputs and preconditions is not possible. So, instead of doing the exhaustive testing we can use risks and priorities to focus testing efforts. For example: In an application in one screen there are 15 input fields, each having 5 possible values, then to test all the valid combinations you would need 30 517 578 125 i.e. (5^{15}) tests. This is very unlikely that the project timescales would allow for this number of tests. So, assessing and managing risk is one of the most important activities and reason for testing in any project. As Exhaustive testing is impossible we only test our system based on the risks, critical modules, timeframe we have and budget i.e. is called optimal Testing.

3) **Early testing:** In the software development life cycle testing activities should start as early as possible and should be focused on defined objectives. This states that Testing phase should be started as early as possible from the requirements gathering phase in SDLC by analyzing the SRS, BRS documents a tester can get basic idea on user requirements and specifications before involving in to real testing.

4) **Defect clustering:** A small number of modules contains most of the defects discovered during pre-release testing or shows the most operational failures.

Defect clustering is based on the Pareto principle that is 80-20 rule. It can be stated that approximately 80 per cent of the problems are caused by 20 per cent of the modules.

When we are testing new software against the user requirements, defects will be finding at a large numbers in certain flow or area of the code which is more complex or critical. When the same software or code is to be tested again & again, after some changes or modifications, we will find the defects in initial few iterations of testing by identifying & concentrating on the 'Hot spot'.

But after certain number of iterations of testing, as the testing improves we see the defect numbers dropping, the most of the bugs will be fixed & the 'Hot spot' area will be cleaned up.

5) Pesticide paradox: If the same kinds of tests are repeated again and again, eventually the same set of test cases will no longer be able to find any new bugs. To overcome this "Pesticide Paradox", it is really very important to review the test cases regularly and new and different tests need to be written to exercise different parts of the software or system to potentially find more defects. This principle states that the more you test the software, the more immune it becomes to your test-just as insects eventually build up resistance and pesticide no longer works. So, it is suggested for a tester to leave such module and concentrate on risk high modules for finding defects instead of working on the same module till the end.

6) Testing is context depending: Testing is basically context dependent. Different kinds of sites are tested differently. For example, safety – critical software is tested differently from an e-commerce site. This state that each application has different strategy of testing to be followed and modules to be concentrated for examples in banking applications we need to concentrate more on Fund transferring and Transaction modules than usability, GUI where as a common application like Maax.co.nz where it is used by common people highly we need to look on usability and GUI too along with the functionality.

7) Absence – of – errors fallacy: If the system built is unusable and does not fulfill the user's needs and expectations then finding and fixing defects does not help. This states that if the system doesn't meet user expectations and requirements then it is time waste process to carry out Test planning and writing Test cases and scenarios. It's better to work on the requirements, coding again to meet the customer requirements correctly.

1.6 Limitations of software testing:

Limitations are principles that limit the extent of something. Testing also has some limitations that should be taken into mind to set realistic expectations about its benefits. In spite of being most dominant verification technique, software testing has following limitations:

- Testing can be used to show the presence of errors, but never to show their absence. It can only identify the known issue or errors. It gives no idea about defects still uncovered. Testing cannot guarantee that the system under test is error free. There no software product that is error free or bug free in the world.
- Testing provides to help when we have to make a decision to either "release the product with errors to meet the deadline" or to "release the product late compromising the deadline".
- Testing cannot establish that a product functions properly under all conditions but can only establish that product does not function properly under specific conditions.
- Software testing does not help in finding root causes which resulted in injection of defects in the first place. Locating root cause of failures can help us in preventing injection of such faults in future.

1.7 How much Testing is enough?

It is possible to do enough testing but determining the how much is enough is difficult. Simply doing what is planned is not sufficient since it leaves the question as to how much should be planned.

What is enough testing can only be confirmed by assessing the results of testing. If higher number of defects are found with a set of planned tests it is likely that more tests will be required to assure that the required level of software quality is achieved.

On the other hand, if very few faults are found with the planned set of tests, then (providing the planned tests can be confirmed as being of good quality) no more tests will be required.

Saying that enough testing is done when the customers or end-users are happy is a bit late, even though it is a good measure of the success of testing. However, this may not be the best test stopping criteria to use if you have very demanding end-users who are never happy!

- If you have just found a defect, this is a signal to keep testing. It may seem counter-intuitive but in general the more defects you find the more likely it is that there are additional defects.
- If you have only exercised a small portion of the overall functionality and have found defects, then this is a signal to continue testing.
- If you have been testing a particular piece of functionality for a while and are not finding any new defects, then this is a signal for you to stop testing. How much of the system's functionality have you tested?
- If there are significant features that are mostly or entirely not tested, then you will likely not be prepared to recommend it is good to go.
- False confidence is a very real danger. This is especially true for developers. Developers state that if their code compiles, it is good enough to be promoted to acceptance testing which is very dangerous.

Finally, deciding how much testing is enough should take account of the level of risk, including technical, safety, and business risks, and project constraints such as time and budget.

1.8 The Psychology of Software Testing:

A software tester is any human being who tests the software. But of course Software Testing is not as easy as you might think of. Needless to say, it's one of the most important and difficult tasks in the entire gamut of software development life cycle process. Nevertheless, the quality of the job done by the software tester is directly proportional to his or her psychological maturity and profoundness acquired, adopted and developed with age and experience.

"The psychology of the persons carrying out the testing will have an impact on the testing process [Meyer 1979]."

1.8.1 Role and Characteristics of a software Tester:

Software Development & Software Testing both aim at meeting the predefined requirements and purpose. But the general outlook towards these two individuals' tester and developer is psychological rather than distinctive classification.

Software Testers require technical skills similar to their development counterparts, but software testers need to acquire other skills, as well.

- Keen Observation
- Detective Skills
- Understanding Product as integration of its parts
- Customer Oriented Perspective-Feel like end user while Testing-Positive testing
- Destructive Creativity-Break the application attitude-Negative Testing
- Organized, Flexible & Patience at job
- Objective & Neutral attitude

1.8.2 Tester/Developer Mindsets:

Both Software Developers and Software Testers have a fundamentally different mindset, but ultimately one goal – to deliver top quality products and services that meet or exceed the customers' needs. A developer's job is to build develop the application. A tester's job is to break the application- Opposite ends of the mindsets, yet interdependent. The challenge for developers testing their own work, rather than handing it off to a tester, is the possibility they might overlook errors, forget to make changes, or generally feel too optimistic. Because a tester's job is to look for errors, the work will be scrutinized with more care. Thus developer and testers will have different mindsets.

1.8.3 Self testing/ Independent testing:

The comparison made on the mindset of the tester and the developer in the previous section is just to compare the two different perspectives. It does not mean that the tester cannot be the programmer, or that the programmer cannot be the tester, although they often are separate roles. In fact programmers are the basic testers. They always perform Unit testing to test their components build. However we all know that it is difficult to find our own mistakes. So, programmers, architect, business analyst depend on others to help test their work. This other person might be some other developer from the same team or the Testing specialists or professional testers. Giving applications to the testing specialists or professional testers allows an **independent test of the system**. These are several level of independence in software testing which are listed below from the lowest level of independence to the highest:

- Those who wrote the code.-Same developer
- Members of the same development team.
- Members of a different group (independent test team).
- Members of a different company (a testing consultancy/outsourced).

Of course independence comes at a price it is much more expensive to use a testing consultancy than to test a program oneself.

Testers need to use tact and diplomacy when raising defect reports. Defect reports need to be raised against the software in non personal way, not against the individual who made the mistake. Because testing can be seen as destructive activity we need to take care while reporting our defects and failures as objectively and politely as positively without criticizing anyone. When we raise defects in a constructive way, bad feeling can be avoided.

We all need to focus on good communication, and work on team building. Testers and developers are not opposed, but working together, with the joint target of better quality systems.

Chapter 1 Sample ISTQB Questions:

Q1. Which option is a part of the "Analysis and Design' area of the Fundamental Test Process?

a. Developing the tests.

b. Writing test summary.

c. Comparing actual and expected results.

d. Analyzing lessons learnt for future releases.

Q.2: Arrange these fundamental test processes in correct chronological order?

a. Implementation and execution, planning and control, analysis and design.

b. Analysis and design, Evaluating exit criteria and reporting, Test closure activities.

c. Evaluating Exit criteria and reporting, Implementation and execution, analysis and design.

d. Evaluating exit criteria and reporting, Test closure activities, analysis and design.

Q.3: Which statement is most true?

a. Different testing is needed depending upon the application.

b. All software is tested in similar way.

c. A technique that will not find defects is not useful.

d. No defects found during testing is a proof of correctness.

Q.4: Which of the following statement is NOT correct?

a. Early test Design can find faults.

b. Early test design can prevent fault multiplication.

c. Early test design can cause changes to the requirements.

d. Early Test design takes more effort.

Q.5: What is the most appropriate definition of quality?

a. Zero Defects.

b. Conformance to requirements.

c. Work as designed.

d. Job done

Q.6: In which phase of fundamental test process, a tester reviews the test basis?

a. Planning and control.

b. Analysis and Design.

c. Test Closure activities.

d. Test Implementation and execution.

Q.7: Which of the following is normally not a testing objective?

a. Finding defects

b. Preventing defects.

c. Providing information for decision making.

d. To prove that software is without fault.

Q.8: Major task of which phase of fundamental test process is to report discrepancies?

a. Evaluating test criteria and Reporting.

b. Test closure activities.

c. Test Implementation and execution.

d. Analysis and Design.

Q.9: Which is the major task of test planning?

a. Determining the test approach

b. Preparing test specifications.

c. Evaluating exit criteria and reporting.

d. Measuring and analyzing results.

Q.10: Enough testing is done when:

a. No more faults are found.

b. Running out of time.

c. A required level of confidence has been achieved.

d. User is not able to find any serious fault.

Q.11: Which of the following requirement is testable?

a. system should be portable.

b. system should be user friendly

c. The response time of load of a specific design should be less than 3 seconds.

d. The safety parts of the system should have 0 faults.

Q.12. Most appropriate definition of Failure is:

a. Deviation from the specific behavior.

b. The result of a mistake.

c. A user action that produces incorrect result.

d. An incorrect process in computer program.

Q.13: Comparing actual results with expected results is a part of which Fundamental process?

a. Test Analysis and Design.

b. Test closure activities.

c. Test Implementation and Execution.

d. Evaluating Exit criteria and Reporting.

Q.14: Test cases are designed during:

a. Test Recording.

b. Test Planning.

c. Analysis and design.

d. Test specification.

Q.15: What is NOT correct for a Test Design Technique?

a. A process to Evaluate testability of test basis and objects.

b. Designing the test environment setup.

c. Reviewing the test basis.

d. A way to measure quality of software.

Q.16: How much testing is enough?

A. It depends on the level of risk.

B. One can never answer this question

C. It depends on the experience of the developers.

D. When you stop getting bugs.

Q.17: A Test team is consistently finding a very good percentage of the defects in a system under test. The test manager thinks that it is a good defect detection percentage but, senior management is not happy with the test group as they think that test team is missing lot of bugs. On the other hand, users are happy with the system, which of the testing principle will help the manager to explain the reason behind the defects that are missed?

A. Exhaustive testing is impossible.

B. Defect clustering.

C. Pesticide paradox.

D. Absence –of-errors fallacy.

Q 18.You is a Test facilitator of a software quality meeting. One person has been dominating the meeting. Which of the following techniques should you use to get other team members into the discussion?

a. Change the topic in which that person does not have strong opinion.

b. Express a different opinion than that person in order to encourage others to express their views.

c. Let the person pause, acknowledge his opinion and then should ask other's point of view on same topic.

d. Confront the person and ask that other team members be allowed to express their opinions.

Q 19 A developer is working on a very complex code. Which of the following Testing principle may affect his work?

  a.   Pesticide Paradox

  b.   Early Testing

  c.   Defect clustering

  d.   Absence of error fallacy

Q20. IEEE standard for Software Test Documentation is?

  a.   IEEE 829

  b.   IEEE 828

  c.   IEEE 827

  d.   IEEE 825

Notes:

2. Testing throughout the software life cycle

2.1 Introduction

Before learning software Testing we have to learn the basic concepts of software engineering and how the software systems are designed, developed and maintained. In this section we are going to present you a clear picture and understanding of how the software engineering process is carried out in companies and their approaches from initial phase to deployment phase. We will describe and discuss the basic terminologies associated with software engineering. As a QA analyst we should have a clear understanding on how the process of software development and deployment takes place to Test it.

2.1.1 Software Engineering:

The application of a systematic, disciplined, quantifiable approach to the development, operation and maintenance of software is called software engineering. The term software engineering is popularized by F.L. Bauer during NATO software engineering conference in 1968.

It encompasses techniques and procedures, often regulated by software development process, with the purpose of improving reliability and maintainability of software systems. The discipline of software engineering which includes Knowledge Tools, and methods for software requirements gathering, software designing, software construction (coding, development), software testing, software deployment and Maintenance of those developed and delivered software's for future updates or for failures. Software engineering is embedded in all kinds of systems like Transportation – AT HOP cards, Flight Booking system. Medical software like Hospital management system, ECG, MRI scans etc. The other disciplines includes Banking Applications, Insurance systems, Educational Applications, Library Management systems, Telecommunications, Office products and software's, Military and Entertainment etc almost the list is endless. That software application affects nearly every aspect of our daily lives and has become pervasive in our commerce, our culture and our daily activities.

2.1.2 Common problems in software development:

1. Poor requirements: If requirements are incomplete, too general, or not testable there will be problems.

2. Unrealistic schedule: If too much work is crammed in too little time, problems are inevitable to meet the delivery time.

3. Inadequate testing: No one will know whether or not program is any good until the customer complaints or system crash. Insufficient time for testing due to delay from developers or lack of proper planning.

4. Miscommunications: If developers do not know what's needed or customers have enormous expectations, problems are guaranteed.

Hence this shows us there is a need of developing and delivering those various kinds of software applications either products or projects with extra care and confidence to the customers. As a Test engineer we must take care of building confidence to the software with thorough Testing and assures quality to the products and projects before delivering them to the customers.

Those software applications developed all over the world may be either a project or a product. So, now let us find difference between a product and project.

Product: If something is developed based upon the company specifications and standards used by a specific group or by many people is called product. Product development is done assuming a wide range of customers and their needs. This type of development involves customers from all domains and collecting requirements from many different environments. Examples of products are Microsoft office, CRM packages, ERP tools, visual studio, Eclipse etc.

Project: If something is developed based on the client requirement and client specifications to meet his goals and to serve his business and to his customers is called a project. Usually projects will be initiated from the customers quote by a kick off meeting with CEO of software developing company. Examples of projects are desktop, web Applications and Mobile Applications of any company like with a specific brand name and are developed to serve their specific group of customers of that brand.

2.2 Software Development Life Cycle Models (SDLC):

A software development life cycle is the process followed to develop new software from the initial stage of information gathering, all the way through maintenance and support of the system. An SDLC should result in a high quality system that meets or exceeds customer expectations, within time and cost estimates, works effectively and efficiently in the current and future planned information technology infrastructure and is cheap to maintain and cost-effective to enhance and upgrade.

SDLC is composed in to the several phases, each comprised of multiple steps. Any software Development has to go through the following steps:

- Business Requirements Gathering
- System Requirements Gathering
- Design
- Coding or Development
- Testing
- Release and Maintenance

2.2.1 Business Requirements Gathering:

Every software development will be initiated with this phase where customer requirements, specifications and expectations for which software to be developed are gathered. The stakeholders or the customer clients are interviewed in the initial phase to gather information in the initial phase which may have following questions. Usually Business Analysts (BA) people will gather all these information from customer and prepares Business requirements specifications (B.R.S.) document.

1. Business vision, Strategy, Goals and objectives to be achieved by the software being developed.
2. Key Business issues and their impact.
3. Current problem
4. Adaptability to the future market needs
5. Financial impact and Budget constraints

2.2.2 System Requirements gathering:

Based on the Business requirements gathered, further analysis is carried out to gather their system related expectations like technical and functional specifications. Usually System Analyst (S.A.) in companies gathers these following requirements from the customers and prepares SRS system requirement specification documents.

1. No of users should access the software being developed per day, per hour and any specific days like Friday nights, peak hours depending on the nature of system.
2. For web applications- Is there need of session maintenance in terms of memory database requirements?
3. What are the browser compatibilities of the application that is being developed? like IE, Chrome, Safari, Mozilla etc?
4. Does this application need to deal with International currencies?
5. Does this application need to deal with multiple languages?
6. Does this application need to be work with Mobile phones and Tablet phones?

2.2.3. Design:

By studying those SRS and BRS documents from the Business Analyst and System Analyst in the company Designers will start designing a plan for the solution. It includes low level component to high level component designs like look and feel of the application and user friendliness UI/UX issues. In this phase of design software overall structure and its nuances are defined. In terms of client server technology, the no of tiers needed , package architecture and database design etc. Technical Analysts (T.A.) will provide high level and low level designs to the designers, developers to start designing and development.

2.2.4. Development or Coding:

Once the basic designs are finished from the designers by clearly understanding the customer requirements from the SRS, BRS documents developers will start converting the high level and low level designs from Technical Analyst in to machine readable format by using any specific technology like .Net, Java, PHP etc. Program tools like compilers, Interpreters and debuggers are also used in this phase of code generation. Developers will generate the code by module wise in a step by step approach.

Designing and coding are considered as most crucial phases of Software development Life cycles any glitches found in the design phase will cost a huge amount in the later stages.

Based on those designs and SRS, BRS documents developers will start coding of the application level by level i.e. module by module.

2.2.5. Testing:

 As a Test Analyst our job starts from here once the SRS and BRS documents are base lined the task of testing is started by reading and analyzing those documents. Separate group of software testers within the company will test the application with respective to the customer specifications and will find errors. Defects/flaws/bugs/errors found at this stage will be sent back to the developer for a fix and have to be re-tested. This phase is iterative as long as the bugs are fixed to meet the requirements of customer before releasing the software.

2.2.6 Release and Maintenance:

In the software life cycle, the release and maintenance phase is the last stage of the cycle. After software passes the design stage and is implemented, tested and released to the customer. The maintenance phase of the software life cycle begins. Understanding the characteristics of the maintenance phase of the software life cycle allows individuals tasked with analyzing the performance of the software after deployment to correctly resolve issues that arise after releasing the software. The Maintenance Phase occurs once the system is operational. It includes implementation of changes that software might undergo over a period of time, or implementation of new requirements after the software is deployed at the customer location. The maintenance phase also includes handling the residual errors that may exist in the software even after the testing phase called failures. This phase also monitors system performance, fixes bugs and requested changes and updates to meet the current technological needs of a customer. For example Initially Banking application ABC is developed, tested and deployed to use it in the desktops and laptops later they felt they need their application to be working even from Mobile phones to meet their customer needs which is called an update .Usually Junior technical professionals both QA and Developers in a CCB (Change Control Board) team are responsible for future updates and maintenance by conducting Impact analysis for updates and root cause analysis for Failures that occur in this phase.

2.3 Various Types of SDLC Models:

The way of your approach to a particular application for testing greatly depends on the life cycle model it follows. This is because, each life cycle model places emphasis on different aspects of the software i.e. certain models provide good scope and time for testing whereas some others don't. So, the number of test cases developed, features covered, time spent on each issue depends on the life cycle model the application follows.

No matter what the life cycle model is, every application undergoes the same phases described above as its life cycle.

Following are a few software life cycle models, their advantages and disadvantages

2.3.1 Waterfall Model:

The Waterfall Model illustrates the software development process in a linear sequential flow. This means that any phase in the development process begins only if the previous phase is complete. The waterfall approach does not define the process to go back to the previous phase to handle changes in requirement. Therefore, different projects may follow different approaches to handle such situations. The waterfall approach is the earliest approach that was used for software development. Initially, most projects followed the waterfall approach because they did not focus on changing requirements. This model is also called as LINEAR SEQUENTIAL MODEL as the entire process done in this model in a linear way. This is the most traditional model in SDLC and this model is selected when customer requirements are clear and no changes required in future. Water fall model is not suitable in Dynamic environments.

Below diagram of water fall model shows the task of each person involved in the each phase from the beginning Business and System Analyst Will involve in requirements gathering and prepare SRS and BRS documents than the as shown in the diagram it follows till the Maintenance phase. No phase will be started without completing the previous stage in the waterfall model.

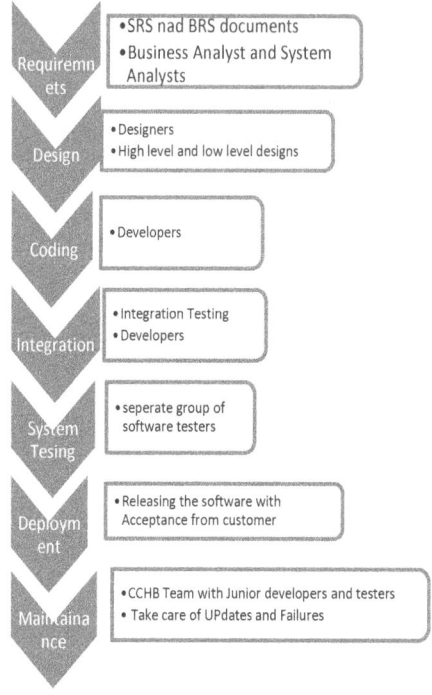

Figure: 1.3.1 **Water Fall Model**

Advantages:

- Provides quality gates at each phase of life cycle

- Simple and easy to implement

- Works well for the small projects where requirements are clear

Disadvantages:

- Poor model for long and ongoing projects

- Time consuming

- High amount of risks and Uncertainty

- Poor model for where requirements are moderate to change

2.3.2 Incremental Model:

In incremental model the whole requirement is divided into various builds. Multiple development cycles take place here, making the life cycle a **"multi-waterfall"** cycle. Cycles are divided up into smaller, more easily managed modules. Each module passes through the requirements, design, implementation and **testing** phases. **A working version of software is produced during the first iteration, so you can test the software early during the software life cycle. Subsequent iterations build on the initial software produced during the first iteration.** Each subsequent release of the module adds function to the previous release. The process continues till the complete system is achieved.

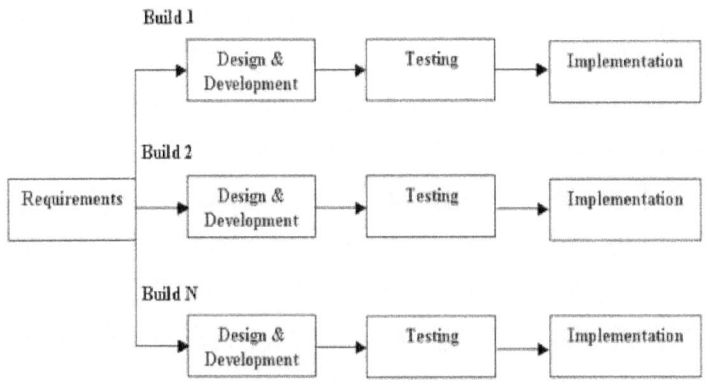

Incremental Life Cycle Model

Advantages:

- Generates working software quickly and early during the software life cycle.

- Easier to manage risk because risky pieces are identified and handled during each iteration

- Customer can respond to each build.

Disadvantages:

- Needs good planning and design.
- Needs a clear and complete definition of the whole system before it can be broken down and built incrementally.
- Total cost is higher than waterfall model.

Incremental model is used when the

- Requirements of the complete system are clearly defined and understood.
- Major requirements must be defined however; some details can evolve with time.

2.3.3. Prototype Model:

Prototype is a working model of software with some limited functionality. Prototyping approach, also known as evolutionary approach, came to picture because of failures that occurred in the final version of the software application developed using the waterfall approach. Failure occurs to change of requirements in future makes the need of re doing the software. In the requirements phase corresponding Business Analyst of companies will show demo through away prototype screenshots to the customer to get his approval. For example in Banking application BA will show various applications format and Menu types Features to the customer in the requirements phase to get his approval. If the customer does not approve the prototype, the development team revisits with the new prototype and resubmits it to the customer for approval. This process continues until the prototype is approved. By using this prototype, the client can get an "actual feel" of the system, since the interactions with prototype can enable the client to better understand the requirements of the desired system. Prototyping is an attractive idea for complicated and large systems for which there is no manual process or existing system to help determining the requirements.

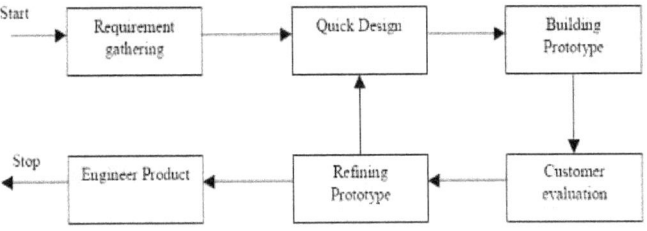

Figure : Prototype Model

Advantages:

- Improved and increased user involvement and we can get early feedback from the customer.

- Reduced time and costs.

- Quicker user feedback is available leading to better solutions

- Confusing and difficult modules can be easily identified in the early stages and extra care can be taken.

Disadvantages:

- The effort invested in building prototypes may be too high if not monitored properly.

- Excessive development time for prototypes.

These prototype model is used when the customer requirements are not clear and when the developing systems needs a lot of user interaction and feedback we will use prototype model.

2.3.4 Spiral Model:

Spiral model is similar to Incremental model, with more significance on Analysis phase. The spiral model has four phases: Planning, Risk Analysis, Engineering and Evaluation. A software project repeatedly passes through these phases in iterations (called Spirals in this model). The baseline spirals, starting in the planning phase, requirements are gathered and risk is assessed. Each subsequent spiral builds on the baseline spiral. Requirements are gathered during the planning phase. In the risk analysis phase, a process is undertaken to identify risk and alternate solutions. A prototype is produced at the end of the risk analysis phase.

Software is produced in the engineering phase, along with **testing** at the end of the phase. The evaluation phase allows the customer to evaluate the output of the project to date before the project continues to the next spiral (iteration).

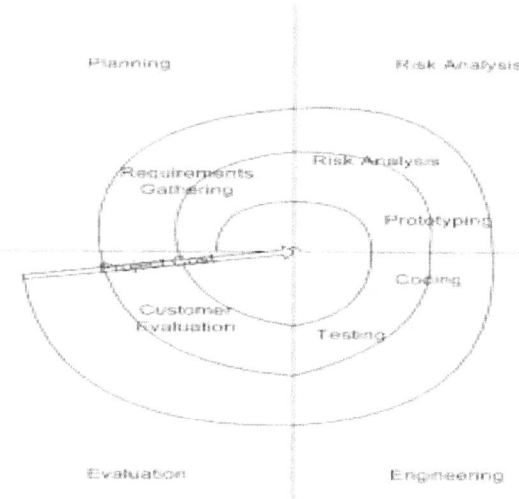

Figure 1.3.4 spiral model

Spiral model is very expensive model and can be used for developing Machine critical software applications. Software is produced in the early phases of life cycle and any updates to the developed software can be added in the later stages which are flexible.

Spiral doesn't work for small projects and needs high skills expertise in the risk analysis phase where the success of the software is truly dependent in this spiral model. Spiral model is an expensive model.

2.3.5 Fish Model:

This model resembles with waterfall model. It involves in continuous/ simultaneous deployment verification and validation processes. Fish model involves more verification and validation in each every phase from the initiation phases like requirement analysis a separate tester groups will review the Requirements documents and designs and coding phases. At each stage a separate testing team will be recruited to for Testing.

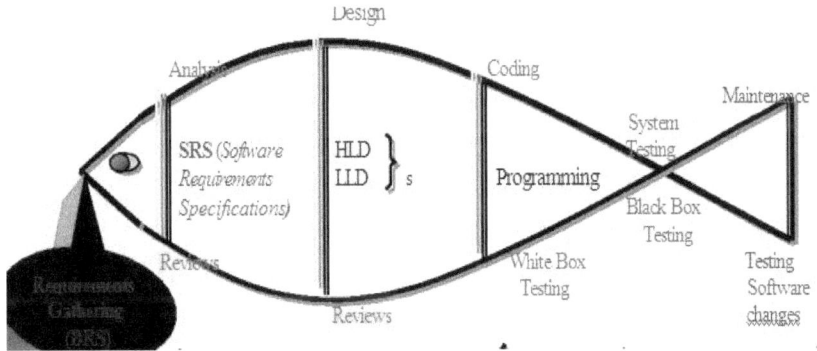

1.3.5 Fish Model

By using the fish model due to phase wise testing we can achieve the best quality product with less error rate in it. But it is very time consuming and very costly affair to adapt fish model will be used only for Machine critical projects such as NASA, Artificial intelligence, Robotic projects.

2.3.6 V –Model:

V model is considered as the Advance SDLC models proposed William Evans ferry where there is mapping between development and testing. V model is the most popular model in SDLC models which is also called as VEE model or Validation and verification model. In V model the approach will be like multiple stages of development and multiple stages of testing which means both testing and development will be started from the initial phase once BRS and SRS documents are baseline both the Testers team and Developers team will involve in to the work. Before the development started the system Test plan is created by the Test lead by analyzing BRS, SRS documents. The Test plan will focus meeting the requirements of specifications in requirements gathering. High level design phase focuses on system architecture and design where Integration test plan is created simultaneously and tested the integration of individual modules units in to single module. Likewise Testers will involve from the initial phase of development life cycle in V model.

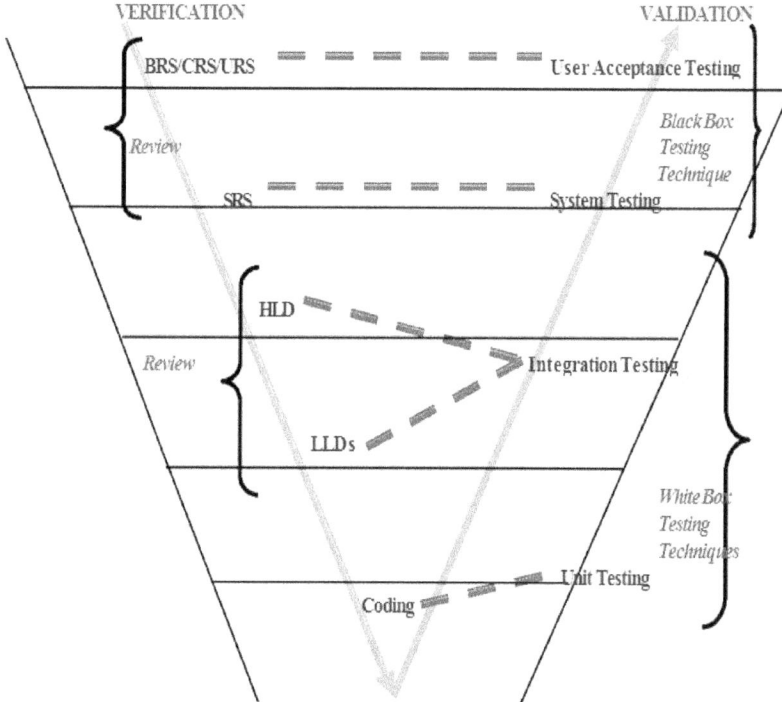

Advantages:

- Easy to use and implement

- Each stage has specific deliverables

- Higher chance of making a quality product

- Time saving

Disadvantages:

- Model doesn't provide clear path for defects found in Testing Phases in later stages.

2.3.7 Agile Model:

Agile software development is a methodology for undertaking software development projects in which incremental functionality is released in smaller cycles, and work is performed in a highly collaborative manner by self-organizing teams that embrace and adapt changes to ensure that customer's needs are truly met. Agile Software Development is not new, in fact it was introduced in the 1990s as a way to reduce costs, minimize risks and ensure that the final product is truly what customers requested.

The idea behind the Agile approach is that instead of building a release that is huge in functionality (and often late to market), an organization would adapt to dynamic changing conditions by breaking a release into smaller shorter cycles of 1 to 6 weeks. Each cycle is called an iteration, or sprint, and it's almost like a miniature software project of its own, because it includes all of the tasks necessary to release the incremental new functionality. In theory, at the end of each sprint, the product should be ready for a QA release. Agile methodology emphasizes real-time communication, preferably face-to-face, versus written documents and rigid processes. In addition, one of the most broadly applicable techniques introduced by the agile processes is to express product requirements in the form of user stories. Each user story has various fields including an "actor", a "goal" or task that they need to perform, an explanation.

In Agile model usually a person with technical knowledge who represents the customer will act as a project coordinator where Test lead and Team lead will contact the customer representative on daily or weekly basis to let him know the status of their application under testing or development by showing him the sprints (models). The testing terminologies will also change in this agile model like

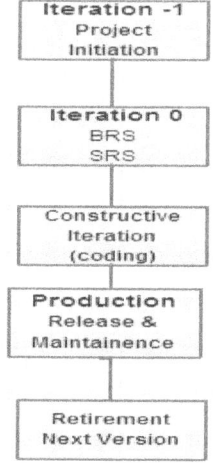

Figure 1.3.7 : Basic Agile Model terminologies

Most agile teams include all the people necessary to release software. At a minimum, this includes programmers and the group or team they are developing the application for, often referred to as their "customers" (customers are the people who define the product; they may be product managers, business analysts, or actual customers). Typically an agile team will also include a Scrum Master, testers, interaction designers, technical writers, and managers. A **Scrum Master** is like a traditional project manager in the sense that he/she oversees the centralization of team communication, requirements, schedules and progress the project being developed in Agile model.

Agile model follows different methodologies to deploy projects depending on the factors Time, Budget and workforce within the company. Various methodologies in Agile are:

- Adaptive software development (ASD)

- Feature driven development (FDD)

- Crystal clear

- Dynamic Software Development Method (DSDM)

- Rapid Application Development (RAD)

- Scrum Methodology

- Extreme Programming   i.e.   (Agile + prototype model)

- Rational Unify Process (RUP)

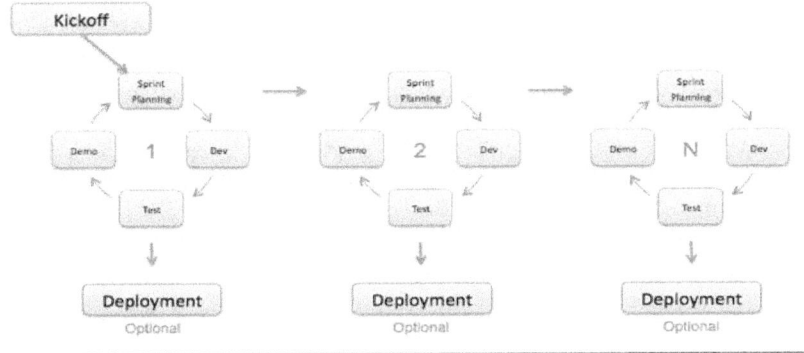

Figure: Agile Model

Agile model is adapted when new changes needs to be made for a project already developed because the freedom agile gives to change is very important. New changes can be implemented at very little cost because of the frequency of new increments that are produced. Agile model is used for Time critical applications for a quick and better quality release.

Advantages:

- Customers, developers and testers will interacts each other constantly which give better idea of the project being developed for the customer.

- We can easily implement the late changes in the requirements at later stages.

- Reduces the time frame of developing and testing compared to other models.

Disadvantages:

- Only experienced programmers and testers are capable to work in Agile models.

- The project can easily get taken off track if the customer representative is not clear what final outcome that they want.

2.4 Test levels: (Levels of Testing)

Levels of testing include the different levels and phases that can be followed from development to deployment while conducting Software Testing. There are four main stages of testing that need to be completed before software can be released for use: unit testing, integration tests, system testing, and acceptance testing.

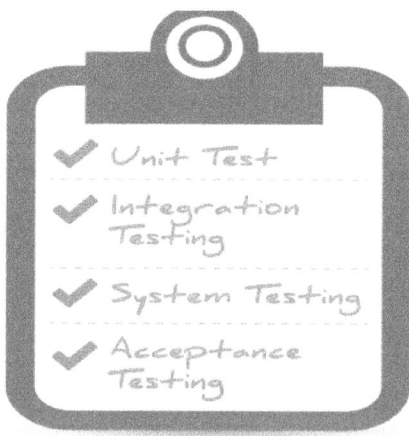

Figure: Levels of Testing

2.4.1 Unit Testing:

In computer programming unit testing is a procedure used to validate the individual units of source code is working properly. A unit is the smallest part of an application in procedural programming terminology a unit may be an individual program, function, procedure etc .Unit testing is also known as component testing or module or program testing is generally performed by programmers who have written the code with a goal of isolating each part of the program and make sure that modules still works correctly. Unit testing helps to eliminate the uncertainty in units themselves then the integration testing becomes much easier. One of the biggest benefits of this testing phase is that it can be run every time a piece of code is changed, allowing issues to be resolved as quickly as possible. It's quite common for software developers to perform unit tests before delivering software to testers for formal testing. In present Technologically advanced generation unit tests are carried by readymade tools and frameworks like Junit etc. Unit testing comprises of functional and Non functional testing characteristics like performance and resource behavior and a bit of structural testing too. Defects found in this unit testing are fixed instantly without any formal recording. unit testing is the cornerstone of Extreme Programming (XP), which relies on automated unit testing framework which can be third party's frame work like xUnit or created within the development team. White box testing techniques are used to perform unit testing.

2.4.2 Integration testing:

Integration testing is the logical extension of unit testing. After unit testing is done all the units are interconnected or integrated and combined as a group to perform testing. It follows unit testing and precedes System testing. Integration testing takes as its input modules that have been unit tested and, groups them in larger aggregates, applies tests defined in integration test plan and delivers it as output the integrated system is ready for system testing.

A typical software project consists of multiple software modules, coded by different programmers. Integration testing focuses on checking data communication amongst these modules. This testing level is designed to find interface defects between the

---

modules/functions. This is particularly beneficial because it determines how efficiently the units are running together. Keep in mind that no matter how efficiently each unit is running, if they aren't properly integrated, it will affect the functionality of the software program. There are different approaches for integration testing. Integration testing can be carried out in different approaches they are as follows:

2.4.2.1 Big Bang Integration Approach:

In this approach all or most of the modules are coupled simultaneously and tested together. The big bang method is very effective in saving time in the integration testing process but, if the test cases and results are not recorded properly, the entire integration process will become more complex. A type of Big bang approach is also known as "Usage Model Testing". In this approach both the hardware elements and software elements are combined in each stage and testing is done.

- Convenient for small projects.
- Fault discovery is difficult.
- Since all the modules are tested once, high critical and complex modules can be easily isolated.

2.4.2.2 Top down Integration Approach:

Top Down testing approach is an incremental integration testing approach which begins by testing the top-level module and progressively adding the low level module one by one. The entire testing carried out in a flow from the main module to the sub module. As we go in to the deeper level we will replace stubs with the lower level modules. Top down approach will use stubs.

STUB: Temporary web page in the place of original page. In the place of under construction sub programs, developers are using temporary programs called stubs.

DRIVER: Temporary web page used in the place of under constructive Main program.

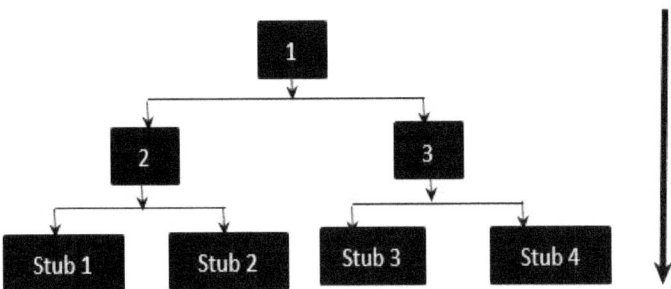

Figure: Top Down Integration Approach

The above diagram clearly shows that Modules 1, 2, 3 are developed ready for integration testing. But, the below modules are under developed pages so stubs are used in those places and integration testing is carried out.

Advantages:

- Provides early working modules of the program so design defects can be found and corrected earlier.

Disadvantages:

- Stubs must be written with utmost care, as they will simulate the total output of the test.

- Requires programming knowledge so developers are suggested for testing in this approach.

- Modules at lower levels are tested inadequately.

2.4.2.3 Bottom Up Approach:

In Bottom Up approach modules at lowest level are developed first and other modules which go towards the main program module are integrated and tested at one time. Bottom Up approach also uses drivers in the place of under construction Main module programs. At the when the code is ready these drivers and stubs are replaced with the actual module.

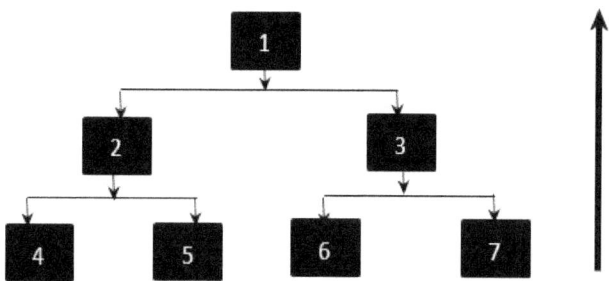

Figure: Bottom Up Integration approach

The order of integration in this approach will be like this:

(4,2) ; (5,2) ;(6,3) ;(7,3) ;(2,1) ; &(3,1).

Though Top level components are the most important, yet tested last using this strategy. In Bottom-up approach, the Components 2 and 3 are replaced by drivers while testing components 4,5,6,7. They are generally more complex than stubs.

Advantages:

- Ideal for applications where Bottom Up methodology is used for designing.

- Test conditions are easier to create.

- Observation of Test results is also easier.

Disadvantages:

- Writing test drivers is harder when compared to stubs because developing a sub module is always easier than developing a Main module.

- Not suitable for the products developed in Top Down Approach.

- Program entity doesn't exist until the last level module is developed.

2.4.2.4 Functional Integration approach:

Integration is carried out based on the functions and functionality of the modules with the reference of functional requirements specifications document. Testing is done along functional data and control-flow paths. First, the inputs for functions are integrated in the bottom-up pattern. The outputs for each function are then integrated in the top-down manner. This approach is also known as Hybrid approach or sandwich Approach as the name suggests we use both Bottom Up and Top down Approaches in this model to leverage the benefits of those approaches effectively. In this approach whole system is seen as three layered one with target layer in the middle another layer above the target layer and final layer which will be below the target layer (middle one).The top down approach is used for top most layer and bottom up is used for the lowermost layer. Hence it is also known as Hybrid or Sandwich Approach.

The primary advantage of this approach is the degree of support for early release of limited functionality. It also helps minimize the need for stubs and drivers. The potential weaknesses of this approach are significant, however, in that it can be less systematic than the other two approaches, leading to the need for more regression testing.

These three Approaches in integration testing apart from the Big Bang technique rest of all three approaches together known as **Incremental Integration Approaches**.

2.4.3 System Testing:

System testing is carried out after the Integration testing is finished and before the Acceptance testing phases. System testing doesn't require any internal coding techniques to be known for testing and follows Black Box testing techniques for testing the application. Usually separate group of testing team carry out this process. System testing is the testing of behavior of a complete and fully integrated software product based on the software requirements specification (SRS) document. In main focus of this testing is to evaluate Business / Functional / End-user requirements. System testing is an investigatory process where the focus is to have destructive attitude and test not only the design, but also the behavior and even believed customer expectations. It is also intended to test up to and beyond the boundaries defined in the software/Hardware requirements specifications. System testing is the real process where the software testers are involved this team will test the entire project/product based on the functional and Non-functional requirements.

It may include tests based on risks and/or requirement specifications, business process, use cases, or other high level descriptions of system behavior, interactions with the operating systems, and system resources. This is the first time end to end testing of application on the complete and fully integrated software product before it is launch to the market or before delivering to the customer.

System is categorized in to two ways they are: (discussed in further section "Testing types")

- Functional testing

- Non Functional testing

2.4.4 Acceptance testing:

Acceptance testing is carried out after finding and fixing all the bugs in system testing phase. Acceptance testing requires fully developed and tested software with all the user requirements meted. It involves both testers, developers with their leads project manager and Customer. Acceptance is the level of testing where the system is tested for the acceptability by the customer before accepting the ownership .Acceptance testing objective is to build confidence that the delivered system meets the business requirements of both sponsors and users. The acceptance phase may also acts as the final quality gateway, where any quality defects previously detected may be uncovered.

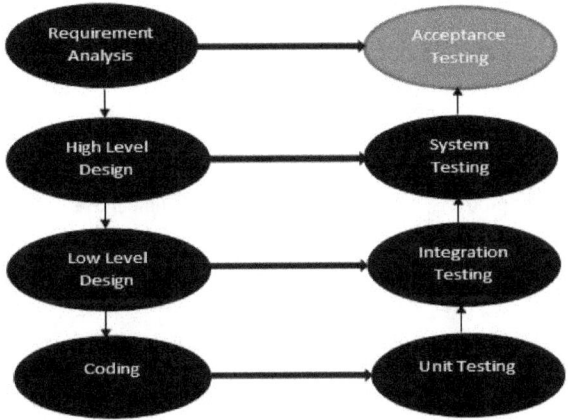

Figure: Acceptance Testing

Upon successful completion of the User Acceptance Testing and resolution of the issues the team generally indicates the acceptance of the application. This step is important in commercial software sales. Once the User "Accept" the Software delivered they indicate that the software meets their requirements. The customers now confident of the software solution delivered and the vendor can be paid for the same.

Acceptance Testing is again divided in to two methods they are:

Alpha Testing:

The alpha testing is the first part of testing. The software needs to pass alpha testing, in order to move on to beta testing. Alpha testing is performed by the users within the organization developing the software. It is done in a lab environment so that user actions can be measured and analyzed. Its purpose is to measure real users' abilities to use and navigate the software before it can be released to the general public. Real customers are invited to the company and collecting feedback from them for their projects being developed and showing them the demo for their project. Alpha testing is suitable for Projects.

Beta testing:

Beta testing is ideal for products, generally involves a limited number of external users. At this time, beta trail versions of software are distributed to a select group of external users, in order to get the public opinion before launching the product. This is done to ensure that the product has few faults or bugs and that it can handle normal usage by its intended audience. If the users find any bugs or faults, they report it back to the developers, who then recreate the problem and fix it before the release. This process helps identify and mitigate defects that were missed during the formal test plan.

Ex; browsers like Chrome and Mozilla releases their beta versions in to public to collect their feedback before releasing their final product in to the market.

 All the testing's (Unit, Integration, System and Acceptance testing) discussed in this section are also called as Dynamic testing techniques which are used in the validation phase.

2.5 Test Types:

To clearly define the objective of a certain level for a program or project Testing Types are introduced. A test type is focused on a specific test objective, which could be the testing of the function to be performed by the component or system; a non-functional quality characteristics, such as reliability or usability; the structure or architecture of the component or system; or related to changes, i.e. confirming that defects have been fixed (confirmation testing or retesting) and looking for unintended changes (regression testing). Depending on its objectives, testing will be organized differently. Hence there are four software test types:

2.5.1 Functional Testing (functions, functionality testing):

Functional testing follows ISO 9126 standards and tests how well (if at all) the system executes its functions. These include any user commands, data manipulation, searches and business processes, user screens, and integrations, compliance, security, accuracy etc.

Functional testing is done using the functional specifications provided by the client or by using the design specifications like use cases provided by the design team.

In functional testing basically the testing of the functions of component or system is done. It refers to activities that verify a specific action or function of the code. Functional test tends to answer the questions like "can the user perform this" or "does this particular feature work". Functional Testing validates the behavior of system.

Functional testing can be carried out in two perspectives.

- **Requirement-based testing:** In this type of testing based on the requirements Tests are prioritized depending on the risk criteria. This will ensure that the most important and most critical tests are included in the testing effort.

- **Business-process-based testing:** In this type of testing the scenarios involved in the day-to-day business use of the system are used. It uses the knowledge of the business processes in daily life. For example, a train journey may have the process along the lines EX for Train journey: search for Train timings, reach the Train station according to time,

Use AT Hop card or purchase paper ticket, Board the Train and alight at your destination.

Steps involved in functional testing:

1. Identify the functions to be tested from the software

2. Create input data based on the specifications document

3. Determine the outputs.

4. Execute the Test cases.

5. Compare the expected output and actual output after execution and determine whether the software is following the user requirements or not.

Functional testing uses the black box testing techniques extensively in its process. The testing types under the functional testing are:

GUI Testing:

During this GUI test, testing team will operate on the AUT (application under testing) screens one by one to confirm those screens are working without any deadlocks or interrupts. This testing is also called as behavioral testing or control flow testing.

Input Domain Testing:

During this test, testing team can test the size and type of every input field in every screen of the software application under test.

Ex: Testing the name field and its limitations, size etc in a registration form.

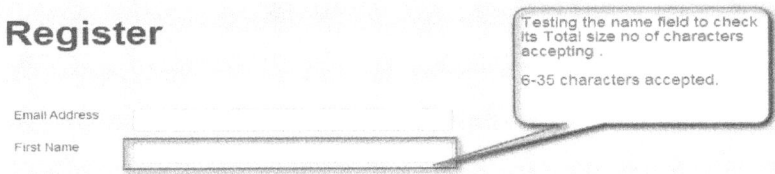

Figure: Input domain testing on Get Skills Registration page.

Error handling testing:

During this test software testing team will operate all the fields in the AUT by giving the invalid inputs and will check for the system behavior when invalid inputs are given. This test is similar to input domain testing but in this test we will test every field with invalid inputs to know how well the system is handling the errors.

EX: NAME field in Get Skills registration page by giving invalid inputs like special characters, integers in Name field instead of alphabets.

First Name

n123242!@#$*|

Please Enter First Name

Output and Manipulation Testing:

During this test testing team will give valid inputs in every field of the software being tested to check and observe the correctness of outputs in every field.

Recovery Testing:

During this test the testing team will operate the AUT application under test by creating illegal operations and observe our software failure tolerance and to check whether our AUT is changing from abnormal to normal state or not. This type of testing is also called as Interruption testing or evil testing.

EX: removing and re establishing the internet connection while using the application, switching off the power supply and observing the application state again after enabling the power supply.

Database testing:

Database testing is a vast concept but during this test our testing team can operate the AUT screens by giving input and will observe the impact of these operations on database. To perform this testing, tester must have good command over SQL commands which are used to communicate with databases.

Data validation is a type of testing where tester's will operate all the fields in AUT and observe for the correctness of given data in screen is stored in the data base or not? And also checks correctness of changes due to data insertion called as Data integrity.

Data Volume testing:

In this test, testers will find the total capacity of our database by inserting the model data in to the database via our application under testing. This testing helps to find the capacity of our application database.

SOA (service oriented architecture) Testing:

During this testing our testers will find the co existence of our application AUT with other supporting software's to share the data. This helps to find out how well our application is communicating with other software applications to share data. This is also called Web services testing and End to end testing or inter system testing.

Simple example for SOA testing is Get Skills application home page contains of YouTube video means our application is using the You tube application's API to display the video in our Get Skills homepage. By clicking the video in our application we will check how well our application is communicating with you tube API these SOA testing has several techniques and readymade tools. Different social media links in our application can also be considered for SOA testing.

Figure: Different social media links in Get Skills home page

Figure: YouTube video displaying in Get skills Homepage

2.5.2 Non Functional testing: (characteristic's, Expectations testing):

After completion of Functional testing, testing team will now concentrate on Non functional testing of AUT- application under test. This testing is also known as characteristics testing because the testing team will now concentrate on characters of the AUT after the functional testing. This testing is also known as expectations testing because every customer will expect usability, accessibility, affordability, performance etc where these aspects are being tested in Non functional testing.

Non-functional testing is concerned with the non-functional requirements and is designed specifically to evaluate the readiness of a system according to the various criteria which are not covered by functional testing.

For example, in functional testing it might be revealed that the function of inputting data into a set of cells in a database works, however, usability testing (a part of Non Functional Testing) reveals that saving a version of the document requires 2 minutes, which will (or should) be judged as too long to wait.

Essentially non-functional testing lets us measure and compare the results of testing the non-functional attributes of software systems, for example, by testing the application or system against the client's requirement or a performance requirement.

So basically, non-functional testing demonstrates how well the product behaves as opposed to simply what the product does.

Of course, while functionality is important, remember that users' affection and trust of a software or system is swayed and affected by non-functional qualities, so one must never forget that non-functional testing is indispensable in its own right.

The universal scenario which tells the importance of Non functional aspects in a software is: Unix is more powerful and capable than Windows OS but more people are using windows OS in the world than Unix because windows concentrates on Look and feel provided usability and ease of use when compared to Unix which is harder to use for non technical person. So for the success of any product both functional and non functional aspects are equally important.

The types of testing that comes under Non functional testing are:

Usability testing:

During this test testing team will observe the user friendliness of Application under testing AUT. User friendliness means ease of use, look and feel and short navigations as well error messages for wrong actions performed etc. The main goal in this testing is to check whether all types of end users can easily use this AUT or not?

Compatibility testing:

This testing is also called portability testing. During this test testing team will operate our Application under test AUT on customer expected platforms like various operating systems windows XP, sever, vista, windows 7 home, premium and Unix, IOS all versions etc. and checks whether our AUT is operating without any errors on various platforms.

Testing our Application under Testing across various versions of all customers expected browsers like chrome, safari, net gear, Mozilla, internet explorer and opera etc to check its browser compatibility also called cross browser testing.

Testing our application on various mobile screens and Tablet phone screens like Nokia, Samsung, LG, and I Phone etc to test its mobile computability.

There are various online resources available now a day to carry out computability testing in a fast manner like: www. browsershots.org , www.crossbrowsertesting.com, Perfecto mobile tool for mobile compatibility testing. Browsers plug-in to check browser computability like IE net render and Pingdom etc.

Hardware configuration testing:

This testing is also known as Hardware computability testing. During this testing team will operate AUT by using different configurations of hardware's. For example testing team can operate our AUT on various processors from the basic level to advanced level and various RAM memories etc.

Performance testing:

Performance testing is a wide concept which helps to find the capacity of our AUT at different loads of traffic. Testers who are skilled in this performance testing are called performance testers they perform testing by using various tools available in the market like HP Load runner, Apache Jmetre, Load UI etc. these are the tests carried under performance testing.

To execute the AUT in customer expected environment with customer expected load (no of concurrent users) to estimate the processing speed for every request is called **Load testing.**

The execution of AUT in customer expected environment with more than customer expected load to estimate the peak load maximum load is called **Stress testing.**

The execution of AUT under customer expected environment and huge load to estimate the server crashing point is called **Spike testing.**

The execution of AUT under customer expected configuration and customer expected load for ex: 500 concurrent users continuously for a long period of time to identify any memory leakages and downtimes is called **Endurance testing, longevity testing or Durability testing.**

Security Testing:

This testing is also called as penetration testing or pen testing which executes the AUT to find out the security flaws in it and loopholes in the AUT is called security testing which is considered as important testing mainly in Banking app's , Military application's etc.

Here testers will check whether the application is prone for any kind of attacks from hackers like DOS denial of service attack , cross site scripting attacks, SQL injections etc. Testing team will perform Authentication and access control testing to check whether the user is valid or not? And whether the user have permission rights to access important information in the application or not? And also checking whether the user's communication and sensitive information like Passwords and bank account details are encrypted or not? Testers can also check for session management in AUT by enabling and disabling the cookies while using the AUT.

SQL Inject Me and ZAP, Qualys SSL Server Test, Tamper Data (Samurai WTF), WebScarab are some of the security testing tools.

Parallel testing:

Here testing Team will compare our AUT with similar application available in the current market to find which features are missed in our application and which makes their applications best and to know the key areas to be focused in AUT for a success in market. For example we are testing a banking application so compare all the available leading banking applications with our AUT is called parallel testing or comparative testing.

Compliance testing:

Here testing team will check whether our Application developed and tested, is following all our company standards and guidelines or not is called compliance testing.

This also known as conformance testing where testers will check whether the AUT have complied with all the customer specifications and have met all the requirements specified in the SRS and BRS or not?

This type of testing is for greater assurance and usually conducted by the external organizations and will advertise as this product is certified after doing so.

2.5.3 Structural testing:

Structural testing is the process where we concentrate mainly on the internal structure of the AUT. This structural testing is also familiar with the name architecture testing where we will try to observe the internal architecture of AUT in the sense observing the internal coding structures module wise or function wise. Usually structural testing uses white box testing techniques to follow this process. Structural testing process is carried out by programmers usually because this approach needs strong programming skills to test the internal coding structures of AUT. The main objective of this type of testing is to find out what happening inside the system.

This structural testing can occur at any Testing stages from unit testing to acceptance testing but, more often it occurs in the stages of unit testing and component testing. Structural testing helps to find out the code coverage and also helps to build the applications hierarchy in a well manner. For example if there is IF Loop in the AUT the concerned developer who performs structural testing will prepare Test data and operate the AUT on all the possible combinations i.e. if and all the else conditions to test the code coverage if the final product doesn't achieves 100 % code coverage then again additional tests may be required to run those applications again to achieve 100%.

White Box testing techniques like basic path coverage, Mutation coverage and control structure techniques will be used to perform structural testing. We will discuss briefly about these techniques in the next chapter Testing Techniques.

2.5.4 Change related testing (modifications testing or Maintenance Testing)

This testing is carried out by the testers after the system testing is completed and before the test closure testing team will report all the bugs they got in system testing. These bugs are fixed and then a new modified build will be released for every module or may be every bug based on the severity of bugs. Then again the team of testers will concentrate on these new builds to check for whether the reported bugs are fixed or not?, Is there any new bugs arises due to modifications in the software?, and also testing will check for any impacts of these modified modules on other modules or functions in the entire Application under testing.

Testers will concentrate here on few tests based on the changes made .However these tests may comprise any of the tests discussed before which may have functional or non functional testing's. The common testing's which comes under the change related testing are:

Re-Testing (conformance Testing):

During the testing if any error or bug occurs testing team will report the defect to the developers team with all the information regarding the defect like Testing steps to get this defect and probably any screenshot of the defect, module or function the defect occurred. After reporting the defect developers will fix the defects and will release a next version (modified build) by fixing the defects reported earlier by testing team. Now to check whether the bug was fixed or not Testers will perform re testing for conformation. It is important to perform this re-testing again with the same data and within the same environment to ensure whether the bug is fixed or not? If test passes and the corresponding defect can be closed otherwise it should be properly fixed in the next build and testers will assign re-open status to this bug and will again report in the defect report.

But testing alone this defect is not enough because this fix may have introduced unseen errors in its related modules which should be tested in the next stage known as regression testing.

Regression testing:

The main intention of performing regression testing is to check mainly the modified new build has not created any unintended defects with its related modules and functions in the entire software. Here both newly added functionalities and also the corresponding dependent functionalities are tested to check the impactness of fixes over other modules in the entire software. Regression is performed for every change occurred in the software and most of the regression testing process is carried out by using Automation testing tools because it is very tedious process to execute the same tests again and again to find the defects.

Win Runner, Regression Tester and V test are some of the tools for Regression testing but most of the functional testing tools like Selenium, QTP, and Watir also support regression testing by executing all the previously passed test cases and also by executing the newly fixed test cases related functions.

EX: If an Application contains three modules like Admin, Student, and Staff and tester founded a defect in student module i.e. student are not capable to login with credentials to the application and now this defect is reported and developers will fix this issue and release a new build of software with fixes. Now, we have to test login functionality of the student module and also testers will test the other modules Admin and Staff modules login functionalities to check whether the new fix has introduced any unseen errors in its dependent module or not . This process of checking the dependent module to find the impactness of new fixes over the dependent modules is called regression testing.

Hence these are the testing phases and levels of testing as well as the Testing types which are performed generally in any organization at different stages of testing in the process of software testing. To deploy software to the customer the application should go through all these discussed phases of testing like Unit testing, integration and system testing and acceptance testing to ensure that the software being developed is thoroughly tested and is error free.

Chapter 2 Sample ISTQB Questions:

Q 1.Which of the following option describes objectives for test levels within life cycle models?

a. There is no necessity to define objectives in advance.

b. Each level has defined specific objectives.

c. Objectives are same for each level.

d. Objectives should be generic for any level.

Q2. Which of the following is a non-functional quality characteristic?

a. Usability.

b. Maintainability.

c. Security.

d. Reliability.

Q3. Which of the options is true for Testers in case of fixing emergency changes?

a. There is no time to test the change before it goes live.

 b. To run the retest of the defect actually fixed.

 c. To run a regression test of the whole system in case other parts of the system have been adversely affected.

d. Retest the changed area and then use risk assessment to decide if it is required to run a regression test in case other parts of the system have been adversely affected.

Q4. Which of the following is a functional test?

    a.   Checking response time on online railway booking system.

    b.   Checking the effect of high volume traffic in an online booking system.

c. To check how easy is the system to use.

d. Checking the database contents.

Q5. Arc testing is also known as?

a. Branch testing

b. Agile testing

c.   Beta testing

d.   Ad-hoc testing

Q6. What is true about the component testing?

a. Testing usually involves programmer who wrote the code.

b. Testing is done by a tester.

c. It includes testing of only non-functional characteristics.

d. It is not separately testable.

Q7.Which is not a Component testing?

a. Check the memory leaks.

b. Check the decision coverage.

c. Check the branch coverage.

d.   Check the decision tables.

Q8.Which of the following statements is not true?

a. Performance testing can be done during component testing as well as during the testing of whole system.

b. Component integration testing tests the interactions between software components.

c. Finding defects is main focus in acceptance testing.

d. Test environments should be as similar to production environments as possible

Q9. Which is not a type of acceptance testing?

a. Contract

b. Regulation

c. Operational

d. Component –integration

Q 10. Alpha testing is:

a. Performed by developers at the developing organization site.

b. Performed by customers at their own site.

c. Performed by developers at customer's site.

d. Performed at developing organization' site but not by developing team.

Q11. What is true for regression testing?

It should be performed every month.

- a. It should be performed when the environment has changed.
- b. It should be performed whenever asked by Project manager.
- c. It should be performed as often as possible.

Q 12 Maintenance Testing is:

- a. Testing a released system that has been changed.
- b. Is the assessment of system's readiness for deployment
- c. Testing by users to make sure that the system meets business need.
- d. Testing to maintain business advantage.

Q13. Which of the following is NOT part of system testing?

- a. Tests based on risks.
- b. Performance testing
- c. Top-down integration
- d. usability

Q 14.A inexperienced tester is testing World Wide Web sites. What are the key features he should concentrated upon?

a.   Security aspects.

b.   Performance, load, stress tests.

c.   Interaction between html pages

d.   All of the above

Q 15. Which of the following is the component test standard?

a.   IEEE 829

b.   IEEE 610

c.   BS7925-2

d.   BS7925-1

Q 16. A tester wants to review the application user interface and other factors of the application with the people who will be using the application. What test will it be?

a. User Acceptance test

b. Usability test

c. Performance

d. Security

Q 17.A tester wants to test varying workloads to measure and access the performance of the system and to ensure that the system function properly under these different workloads. What test will the tester use?

a.   Load Testing

b.   Integration Testing

c.   System Testing

d.   Usability Testing

Q 18. Repeated testing of already tested program, after modification, to discover any defects introduced or uncovered as a result of change or changes?

a.   Confirmation testing

b.   Regression testing

c.   Functional testing

d.   Re-testing

Q 19. Impact analysis helps to decide:

    a. How much more re-testing should be done.

    b. How many more test cases need to be written?

    c. Different tools to perform regression testing

    d. How much regression testing should be done?

Q 20. Which of the following is not an integration Testing?

  a. Design-based

  b. Big-bang

  c. Bottom-up

  d. Top-down

Notes:

3. Static Testing Techniques:

The Software testing techniques which are used by the Business Analysts, Developers and Testers in the whole process of software testing. This chapter gives a clear overview of the static testing techniques involved in the software testing and tells by whom and when and at which stage of testing these techniques are used. Mainly there are two software testing methodologies called static testing techniques and Dynamic testing techniques.

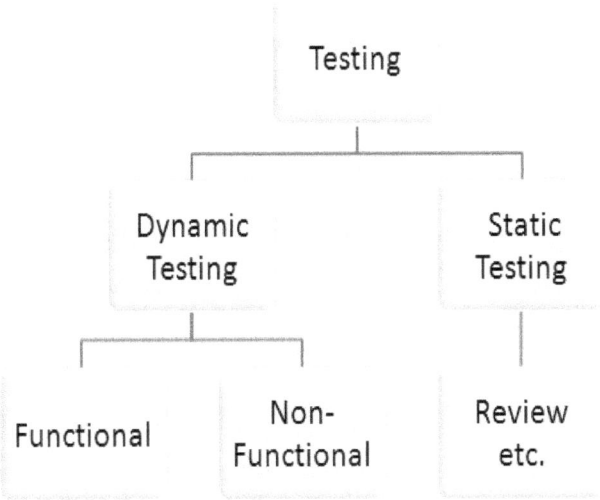

Figure : Static v/s Dynamic Testing

3.1 Difference between Static and Dynamic testing:

Static testing	Dynamic testing
Testing done without executing program	Testing done by executing the program
Verification process	Validation process
Prevention of defects	Finding and fixing the defects
Performed before compilation	Performed after compilation
Cost of finding and fixing defects is low	Finding fixing cost of defects is high
More reviews are recommended for good quality	More defects are highly recommended for good quality
Requires huge number of meetings	Comparatively lesser meetings involved

3.2 Static testing Techniques (Document testing):

These static testing techniques are also known as document testing techniques which is another method in software testing , In dynamic testing we compare the Application under testing by giving inputs and observing actual output to expected output by executing the application under testing. But in this static testing we do not execute the Application under testing we only observe and examine the Application under testing by using tools or manually.

Static testing finds the defects in software rather than failures like in dynamic testing. Static testing techniques which are intended to find earlier issues like missing requirements, design defects, interface problems and specifications missing in software. Project managers, stock holders, business and system Analysts and other interested people may involve in this static testing process by using techniques like Reviews and Walk through.

3.2.1Review:

Review is a process or form of a meeting that could be formal or informal meeting where project heads, customer representatives, business analysts and other interested people examine the Application under testing for a feedback or an approval. Reviews can examine the technical and functional specifications or design, code user documents, Test plan documents, support and maintenance documents and any other type of documents related to the specific Application being developed. Reviews can be conducted at any stages of software development life cycle.

Reviews help to reduce the defect ratio as early as possible prior to development and testing phases. So that reviews can greatly reduces the re working costs and it is also suggested to find defects in the earlier stages to reduce the fixing costs. As a general principle, earlier documents are reviewed; the greater will be impact of its defects on any other follow up activities. Behalf of finding the earlier defects the reviews will acts as informational meetings for customers where they came to know more about their software being developed.

Reviews can be formal or informal. Formal reviews are known as Inspection and informal reviews referred as Walkthroughs.

3.2.2 Inspections:

An inspection is a formal method of review where rigorous in depth group review designed to identify the problems close to the point of origin. The formality of this reviews depends on the maturity of code, legal and regulatory requirements needed etc. Formal reviews are process oriented where the defects are identified but not fixed.

Objectives of inspections are:

- To find defects in the earlier stages of development

- Verifying the AUT meets its requirements

- To build technical knowledge to team people and to provide product status to customer

- To verify the AUT has been presented according to predefined standards

- Increases the effectiveness of software testing.

Roles and Responsibilities in Review process:

According to IEEE 1028 review process contains of the following roles they are:

Inspector leader:

Inspector leader also known as a Moderator is responsible for administrative tasks, planning and preparation of reviews and inspector leader has to ensure that these reviews are meeting their project objectives and he is also responsible for scheduling regular reviews and most importantly collecting the data from the reviews held.

Recorder:

Inspector leader may acts as the recorder. The main responsibility of recorder is to maintain the data of every review held for a better process analysis.

Reader:

Reader will lead the inspection team through the software product in a comprehensive and logical fashion, presenting the sections of work done and highlighting the areas, modules to be focused.

Author:

Author is responsible for deciding the entry criteria of the review meeting and also for the re work status if needed and he will call exit criteria of the process too.

Inspector:

Inspectors are responsible for presenting certain areas in the review meetings. They will describe and explain the defects in software with different viewpoints like requirements, design, safety, code, testing etc. Few inspectors are assigned to review some topics to ensure better quality. For example one inspector may concentrate on GUI and other on security and another on syntax, re usability and maintainability of the software product. The roles and responsibilities of inspectors are assigned by Inspector leader.

Reviews follow the specific process and steps they are:

1. Planning:

Inspector leader perform planning tasks like which products need to be reviewed, identifying the inspection team and assigning roles to them and also checks for the need of training to their inspectors. Inspector leader also identify any materials required for the review process and distributing the materials for stakeholders etc.

2. Kickoff:

Kickoff is an optional step in the process of Reviews but yet it is highly recommended to have a kick off. In this stage all the receivers will receive a short introduction about the documents reviewed and documents yet to be and their relationship with the current reviewed documents. The main objective of this Kick off is to synchronize all the participants to the Review process in depth.

3. Preparation:

All the participants review the documents, rules and standards received from the inspector and will report any issues, questions they have regarding those reviewed documents. All the defects will be recorded individually using a checklist and these annotated forms will be given back to the Inspector lead at the end of the meeting.

4. Review:

The main objective of this review stage is to finalize the defects recorded in the preparation phase by the inspectors. Every defect is categorized depending on its priority like showstopper, high critical, low critical etc. Based on these finalized defects the moderator will now decide to exit the review or to rework on the issues and possibly a decision is made to exit criteria here. If no of defects finded is greater than rework or review status will be assigned by the moderator and again the re-review process is started.

These are the roles and responsibilities and process involved in the formal meeting sessions called Reviews.

3.2.3 Walkthroughs:

Method of conducting informal, individual or group meetings is called walk through. The participants in this walk through process may be designer or a Team leads, Test leads and their members of team. These people may ask their questions and can discuss about the possibility of errors in critical areas and they may discuss about their company standards with the reference of the AUT. Like Reviews Walkthrough is not a formal meet and can be conducted at any time without planning and preparation depending on the necessity. Generally all the people working on the AUT are involved in these Walkthroughs. Author leads the walkthrough process by discussing about the documents or by evaluating his/her work on this project till now and finally gathering their common feedback which helps for the development of better Application. Author presents the artifacts in a step by step manner on which all the audience will discuss and contribute themselves to find errors .Instead of correcting the errors and defect tracking audience will get knowledge on the product in these Walkthroughs.

Goals of Walkthrough:

- Explaining the documents to all the people in the meeting especially for the people who are from Non Technical background.

- Gaining the common understanding to all.

- Discussing about the proposed solutions and thinking for better alternative solutions.

- Focusing on how the product meets all requirements.

Difference between a Walkthrough and Review:

Reviews	Walk Throughs
Formal meeting	Informal meeting
Planned meeting with assigned roles and responsibilities	Unplanned meeting initiated depending on the requirements
Lead by a moderator	Lead by a Author
Recorder records the defects	Author makes note of suggestions and alternatives, feedback.
Reader reads the product documents	Author reads the product documents

These Reviews, Inspections, Walkthroughs are considered as Static techniques or document techniques and used in verification method of Software testing process.

Chapter 3   Sample ISTQB Questions:

Q1. Which option is NOT true for Review?

a.   It is only a manual activity.

b.   Early defect detection and correction.

c.   Reduced cost and time.

d.   Fewer defects and improved communication.

Q2. Arrange these phases of formal Review in a correct order.

I Planning ii. Review Meeting iii. Rework iv. Individual Preparations v. Kick Off vi. Follow Up

a.   ii, iii, iv, v, vi

b.   Vi, I, ii, iii, iv, v

c.   I,v,iv,ii, iii, vi

d.   I, ii, iii, v, iv, vi

Q3. **"Explaining the objectives" is part of which phase of formal review?**
   a.   Planning
   b.   Rework
   c.   Kick-off
   d.   Review meeting

Q4. Who is responsible for allocating time in project schedules and decides on execution of reviews?

a.   Author

b.   Manager

c.   Moderator

d.   Scribe

Q5. Which of the following statements are TRUE for Informal reviews?

      May take form of pair programming.
      The process must be documented.

No formal process.
Inexpensive way to get benefit.

a.   i and iv
b.   I and iii
c.   I , iii and iv
d.   All the options are correct.

Q6. In what phase should static test be used?

a.   Requirement

b.   Design

c.   Coding

d.   All the above

Q7. Which of the following is the main activity of the "Planning" Phase?

a.   Defining the review criteria

b.   Explaining objectives

c.   Noting potential defects.

d.   Gather metrics.

Q8. A scribe is:

a.   The person who leads the review of the document.

b.   The person with chief responsibility for the document to be reviewed.

c.   The person who documents all the issues and problems.

d.   The person who identify and describe findings in the product under review.

Q9. What can static Analysis NOT find?

a.   Array bound violations.

b.   Memory leaks

c.   Dead code

d.   The use of variable before it is defined.

Q 10.What is the most appropriate difference between an Inspection and a walkthrough?

 a. Both are led by Author

b. Both are led by trained moderator.

c. Inspection is led by trained moderator and walkthrough is led by an author.

d. Walkthrough is led by an author and during inspection author is not present.

Q11. Static analysis is best described as:

a.   The analysis of batch programs.

b.   The analysis of program code.

c.   The use of black box testing.

d.   The reviewing of test plan.

Q 12. An Important benefit of inspecting a code is:

a.   Code will be tested before the execution environment is ready.

b.   Are cheap to perform.

c.   Can be performed by anyone.

d.   Are very easy to perform.

Q.13. Which Defect cannot be detected by static analysis tools?

 a. Security vulnerabilities.

 b. Uncalled functions and procedures.

 c. value stored in a variable is correct.

 d. programming standard violations.

Q14. Which of the following is called as "Peer review"?

a.   Walkthrough

b.   Informal Review

c.   Technical Review

d.   Inspection

Q 15. Which of the following are good candidates for manual static testing?

a. Requirement specifications, test plan, code, memory leaks.

b. Requirement specifications, test cases, user guides.

c. Requirement specifications, user guides, performance.

d. Requirement specifications, website, code, use cases.

Q 16. Which of the following is NOT one of the success factors for reviews?

a. Clear objectives for each review.

b. Reduced development timescales.

c. There is an emphasis on process improvement.

d. developer's issues and psychological aspects are not reviewed.

Q 17. Which of the following can be reviewed?

a.   Strategic directions

b.   Project progress

c.   Test plan

d.   All the above

Q 18 What is MOST appropriate for Technical Review?

a.   Widely viewed as useful and cheap. A helpful first step for chaotic organizations.

---

b.   Includes peer and technical experts, no management participation. Normally documented, fault finding.

c.   Author guides the group through a document so all understand the same thing, consensus on changes to make.

d.   Formal individual and group checking, using entry and exit criteria, leader must be trained and certified.

Q 19.Unreachable code would best be found using?

a.   Test management tool

b.   Test Execution tools

c.   Static analysis tool

d.   Coverage tool

Q 20. Which of the following statement is false?

a.   Review help to find faults in development and should be applied early

b.   Static analysis can find faults and give information about code without executing it.

c.   Static analyses do execute the code.

d.   Objective of review is validation and verification against specifications and standards.

Notes:

4. Test Design techniques:

Test Design Techniques are used by the team of software testers to design test cases and Test scenarios to test an application in various ways. The most common Test design techniques used in the software testing are White box testing techniques and black box testing techniques.

4.1 White box testing techniques:

White box testing techniques are limited to unit testing and integration testing phases of testing. White box testing is also known as Glass box testing or open box testing or structure based testing. This name itself tells us where we will concentrate on the internal structure and code of AUT. So, the testers who perform this white box testing need programming knowledge. Usually in most of the companies developers are responsible for conducting Unit and Integration testing which requires usage of White box testing techniques.

Using the white-box testing techniques explained in this chapter, a software engineer can able to Design test cases that (1) exercise independent paths within a module or unit; (2) exercise Logical decisions on both their true and false side; (3) execute loops at their boundaries and Within their operational bounds; and (4) exercise internal data structures to ensure their validity. (Pressman, 2001).

White box testing follows specific techniques in its process of testing to observe the internal coding logic and to exercise looping structures. Static techniques are useful for accessing test coverage and also helpful to design additional test cases. These techniques are as follows:

4.1.1 Basis Path Testing:

Basis path coverage is a white box testing technique used to design Test cases. Basis path testing ensures that every path (positive and negative) throughout the program has been executed at least once. The main goal of this basis path testing is to determine the logical complexity of a procedural design and use this as a guide for defining basic set of execution path.

We design flow graphs which are useful for representing control flow of a process. Any procedural logic can be represented in flow graphs. Basis path testing introduced by James Mc cabe involves 4 steps they are:

1. Compute the program graph.

2. Calculate the cyclomatic complexity.

3. Select a basis set of paths.

4. Generate test cases for each of this path

Flow Graph Notation:

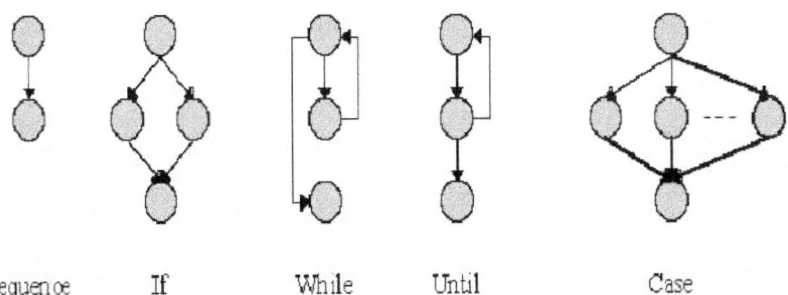

Sequence   If   While   Until   Case

- In flow graphs arrows are called edges which represent flow of control

- Circles are known as nodes which represents actions.

- Areas bounded by edges and nodes called regions.

- Predicate node is a node contains condition. (Sobeys, 1997).

Cyclomatic complexity:

It is software metric which is used to measure the Complexity of the software program. Cyclomatic Complexity metric is based on the number of decision in a program. If a program has two or more decisions like in IF, While, and if else loops then the cyclomatic complexity will be N+1 means there are 3 decisions parts so n=3 and the cyclomatic complexity is 3+1=4.

Cyclomatic complexity is defined as: L – N + 2P, where- L = the number of edges/links in a graph- N = the number of nodes in a graph- P = the number of disconnected parts of the graph (e.g. a called graph or subroutine)   Source: ISTQB Glossary of terms.

Let us see this example for a better understanding of cyclomatic complexity.

 What is the cyclomatic complexity of the following code?

void func1 ()

{

int a,b,c,d;

while (2)

{

if (a>b)

print a

else if (b>c)

print b

else

print c

}

}

Cyclomatic Complexity can be calculated using either of two rule   1) Formal Formula 2) Decision Point rule...

I prefer Decision point rule: Which says CC = Total number of Decision Point + 1.

 Now what is a Decision Point..????

Okay the simplest way is to sum the number of binary decision statements such as For, While, If etc and simply add 1 to it .Hence in the program above, the Cyclomatic Complexity is 3 + 1= 4. (Sather, 2012).

Determining Independent paths:

Independent paths determine the different ways to reach the end node in the flow graph starting from the first node.

Independent paths:

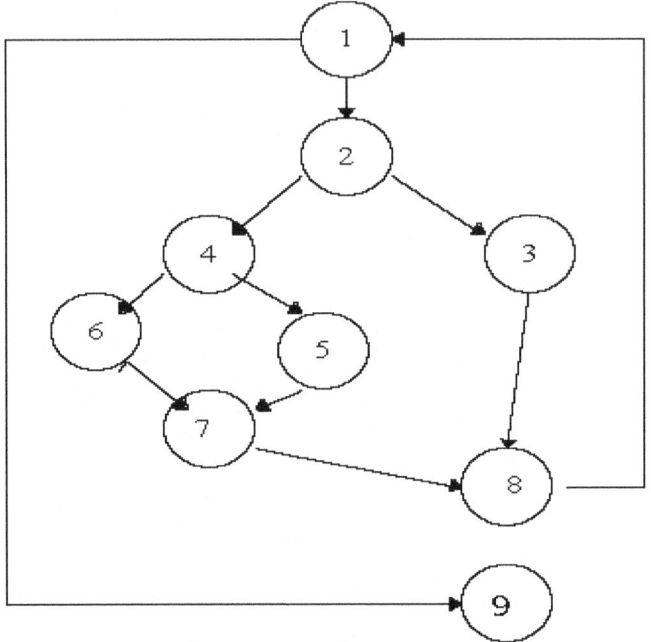

1. 1, 9

2. 1, 2, 3,8,1,9

3. 1, 2,4,5,7,8,1,9

4. 1, 2,4,6,7,8,1,9 (Sobeys, 1997).

After determining the independent paths like this from the above diagram we determined independent paths starting from the first node to reach its end node in all the possible directions. Now, preparing test cases for each path in the basis set will increase test coverage of our application.

4.1.2 Condition Testing:

A condition may be logical expression which gives true or false output. Condition testing is another method in white box testing techniques which allows programmer to determine the path through a program by selectively executing the code by comparing the value against each condition. Here test cases are designed to execute condition outcomes. If two or more conditions joined by logical expression like IF a<b<c then it is called as compound condition.

To check the below if else condition code and to cover all the conditions we need 2 test cases.

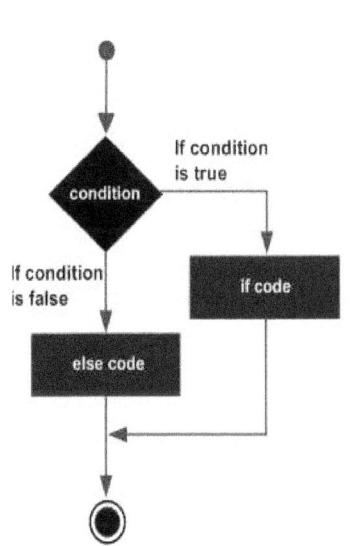

int a =100;

if (a<20)

{

Printf(" a is less");----------Test case 1

}

Else

{

Printf("a is greater than 20");------Test case 2

}

To determine the below switch case we require 3 test cases to cover all the conditions.

Switch (MAL){

Case 0:

Mk=ch-k; --------Test case 1

Break;

Case 1:

Mk=Si*sk;- ---------Test case 2

Break;

Case 2:

Ch=si-mk; ---------Test case 3

Break;

4.1.3 Branch Condition Combination Testing and Coverage:

The Branch condition combination testing needs to execute all the combinations of the attributes i.e. more than two branches. Branch condition combination is a typical testing technique which is very time consuming any risky to perform and requires 2^n test cases to achieve 100% coverage of a condition containing n Boolean operands. For example and conditions contain 3 logical operands namely A, B, C require 8 combinations like:

Case	A	B	C
1	FALSE	FALSE	FALSE
2	TRUE	FALSE	FALSE
3	FALSE	TRUE	FALSE
4	FALSE	FALSE	TRUE
5	TRUE	TRUE	FALSE
6	FALSE	TRUE	TRUE
7	TRUE	FALSE	TRUE
8	TRUE	TRUE	TRUE

This rapidly becomes unachievable task to test more complex conditions.

Code coverage:

To determine what percentage of our applications code is tested and what percentage of code is still yet to be tested we follow code coverage techniques. This is a form of white box testing technique which inspects the code directly.

So, if you have 90% code coverage than it means there is 10% code that is not covered and tested yet. I know you might be thinking that 90% of the code is covered but you have to look from a different angle. What is stopping you to get 100% code coverage?

```
If( Tester. IsPassionateTester())

{

}

Else

{

}
```

Now, in the above code there are two conditions. If you are always hitting the "YES" condition then you are not covering the else part and it will be shown in the Code Coverage results. This is good because now you know that what is not covered and you can write a test to cover the else part.

Just remember, having "100% code-coverage" doesn't mean everything is tested completely - while it means every line of code is tested, it doesn't mean they are tested under every (common) situation.

There are so many tools used for code coverage a good tool will give you not only the percentage of the code that is executed, but also will allow you to drill into the data and see exactly which lines of code were executed during particular test. Ex: Magellan, J Test, Emma, Cobertura, Test explorer etc.

To measure how well the program is exercised by a Test suite, one or more coverage criteria are used. There are no of coverage criteria, the main ones being:

Statement coverage:

Has each line of source code has been executed?

Condition coverage:

Has each condition (true/false) has been executed?

Function coverage:

Has each function in the application is executed?

Entry/Exit coverage:

Has every possible call and return of the function been executed?

Parameter Value Coverage (PVC):

To check if all possible values for a parameter are tested. For example, a string could be any of these commonly: a) null, b) empty, c) whitespace (space, tabs, and new line), d) valid string, e) invalid string, f) single-byte string, g) double-byte string.

Decision coverage:

Has every decision is executed in Not just Boolean expression to be evaluated for true and false once, but to cover all subsequent if-else if-else body.

4.2 Black Box Testing Techniques:

Like white box testing techniques black box testing does not require any programming knowledge and internal coding structure. Black box testing is also known as behavioral testing sometimes which doesn't concentrate on the internal things. Black box testing will concentrate on the external behavioral like 1. Missing functionalities,2. Interfaces, 3. Data structure errors, 4. Performance errors, 5. Validation issues etc.

Usually separate group or independent software testers will perform this black box testing process without checking the internal structures and code like programmers do. Through this testing testers will determine whether the functions are working according to the requirements and will assess their behavior. This black box testing techniques are mainly used in system testing phase to conduct both functional and non-functional testing's. We will always concentrate on the input and outputs without looking in to the code of software. Some of the Black Box testing techniques used by the testing teams are as follows:

4.2.1 Boundary value Analysis (BVA):

BVA technique is most popular techniques in black box testing where we will check the ranges for any input function based on the requirements. Here the range in the sense we will check for the minimum and maximum values that can be fit in to our input functions to know the range. The main conditions to be tested for any input field using BVA technique is:

- Maximum value

- Maximum value + 1

- Maximum value -1

- Minimum value

- Minimum value -1

- Minimum value + 1

Example: let us take an example of Get Skills Web Application to evaluate how the BVA technique works.

Testing steps: 1. Navigate to url:www.getskills.co.nz

2. Click on contact us link in the footers page

3. Perform BVA on the First name field in contact us form.

NOTE: Here the range given for First Name field is (Min1, Max 12) characters

TEST CASE	Characters count	Output
Minimum	1 character	valid
Minimum + 1	2 characters	valid
Minimum -1	0 characters	Invalid
Maximum	12 characters	Valid
Maximum + 1	13 characters	Invalid
Maximum - 1	12 characters	Valid

First Name   a

Figure : First Name field in Get skills Contact us form.

From the above diagram we can check the range of the First name field by entering the characters in to the text box with the rules Min, Min-1, Min + 1 and Max, Max-1 , Max +1 characters to check the range. These BVA test results are tabulated above where two test cases are failed in the conditions Minimum-1 and maximum +1 character. By doing so we can easily evaluate the boundaries of every input field is matched with the requirement specifications or not?

4.2.2 Equivalence class portioning: ECP

ECP is other Black box testing technique which concentrates on the type of input and output conditions entered in to the text box. ECP divides all the input fields in to two categories valid and invalid to check the various test conditions over the input fields. In the BVA example the predefined range is 12 for the input field so we have written 6 Test cases there if there is a case where the range will 120 characters then writing all those 120 test cases is a complex job. To reduce this complexity we only identify the equivalence classes with valid and invalid conditions. Let us look at this again with same example of Get skills First name input field from their contact form.

Valid conditions	Invalid conditions
Lowercase alphabets within range	Blank space
Uppercase Alphabets	Numerics- Integers (1, 22, 9, 38…)
	Special characters (!, &,*,>,<,:,",%)

Based on our previous tests BVA now, we will check for the type of valid input conditions and type of invalid input conditions within the range. So, based on the results Tabulated above our application under testing Get skills - First name field in contact form doesn't accept special characters, Integers and blank spaces in to the input field and since they are marked with red color and stated as Invalid conditions and we can write alphabets in the name field either uppercase and lowercase are accepted with our first name input field so they are tabulated under valid conditions. Thus we can reduce some of the complexity with this ECP technique while testing and we can also know the type of input conditions are valid ones and invalid ones.

4.2.3 Decision table technique:

Decision table technique is good technique to deal with complex projects. Decision table technique gives the mapping between input and output conditions. This technique is also familiar with the name cause effect graph technique. This technique will check different combinations of inputs and outputs and results are clearly tabulated for a visual aid. Decision tables are really helpful to test the complex business logics. These decision tables have two parts namely condition part which contains various conditions and the results part which explains about the actual result obtained with the inputs we have given. In conditions part we will include all the different valid and invalid combinations of inputs to evaluate this decision table.

Let us consider an example for this technique with "successful login" condition from Get Skills web application.

Let us evaluate different combinations of inputs and will check the outputs of Get Skills login screen by using decision table technique.

Test steps:

Navigate to url:www.getskills.co.nz and click on Log in Link at the top right corner.

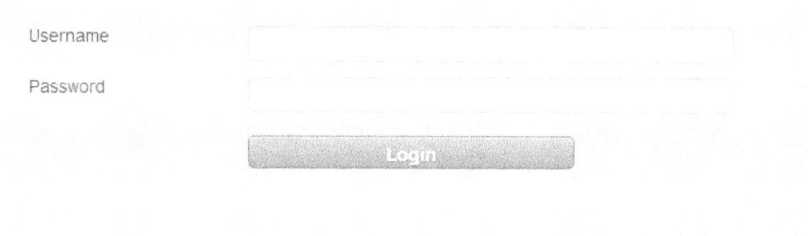

Figure: Get Skills Application Log in screen.

Decision Table for GetSkills Log In:

Username	password	Expected out put
Valid username	Valid password	Successful Login
Invalid username	Valid password	Error page
Valid username	Invalid password	Error page
Invalid username	Invalid password	Error page
Blank username (left user name without filling anything)	Filled with password	Error page
Valid username	Blank password	Error page
Blank username	Blank password	Error page

So, these are the various conditions and combinations of input and outputs we have used to test the applications Log in functionality by using Decision table Technique.

4.2.4 Error Guessing Technique or Experience Based technique:

Error guessing is the last black box testing techniques used in software testing. There no specific tools and strategies to be followed in this error guessing technique. Error guessing comes with an experience with the technology and the project. Error guessing is an art of guessing where errors can be hidden. This is a skill rather than a technique. But to guess errors like these testers will need some experience and some great domain knowledge for example an experienced tester with banking domain can easily guess the errors and missed functionalities in the application.

- Some tips that can help for error guessing are:

- Knowledge on previous executed test cases to discover the most critical modules to concentrate for error guessing

- Experience in testing similar related systems.

- Clear understanding on requirements and user specifications.

Error guessing is a skill set which can make the Testing process more effective and efficient but this is also a skill set which can acquire with great experience in testing. Even though it is not a formal technique in black box method we can document the errors at any stages of testing just like in other black box testing techniques. Sometimes error guessing is also familiar with the name experience based testing.

Chapter 4 Sample ISTQB Questions:

Q 1. What is definition of Test design technique?

a. A procedure for selecting and designing tests based on structural or functional model of the software.

b. It is an event that could be verified by one or more test cases.

c. The sequence of actions for the execution of a test.

d. To assess and design only the functional aspect of the software.

Q2. A wholesaler sells cardboard boxes. He rejects the order if it is less than 10. There will be a discount of 10 % if you order more than 100 boxes. If you have to prepare test cases using different values for the number of boxes ordered. Which of the following option covers all the equivalence partitions?

a. 10, 15, 100

b. 1, 20, 101

c. 5, 10, 99

d.   10, 15, 99

Q3.Which standard for "software test documentation" describes the content of test design technique?

a.   IEEE STD 829

b.   IEEE 610

c.   BS7925-1

d.   BS7925-2

Q 4. Expected results should ideally be defined prior to?

a.   Test planning.

b.   Test Analysis.

c.   Test Execution.

d.   Evaluating exit criteria.

Q5. During which phase are the test cases developed, prioritized and organized in the test procedure specification?

a.   Test Implementation

b.   Test planning

c.   Test analysis and design

d.   Test closure activities

Q. 6 which is a common characteristic of black-box testing?

a. To provide information about the how the software is constructed.

b. Test cases can be derived systematically from these models.

c. The knowledge and experience of people are used to derive test cases.

d. The extent of coverage of software can be measured for existing test cases.

Q 7. Which is the odd one out?

a.   White box

    b.   Glass box

    c.   Structural based

    d.   Specification based.

Q8. In an hospital charges for children under and equal 5 years of age are $ 0, for patients between 5 and 65 years is $ 20 and for patients above 65 years is $10.Identify the equivalence values that belong to same class?

    a. 6, 25, 60

    b. 3, 5, 65

    c. 60, 65,70

    d. 4, 5, 65

Q.9 A system validates data on a person's age, which should be between 1 and 99.By Using boundary value analysis which will be the most appropriate one?

a. 0, 1, 2, 99.

b. 1, 98, 99, 100

c. 0, 1, 99, 100

d. -1, 0, 1 , 99

Q 10. Use case Testing is mainly used in.

    i.   Ensuring that mainstream business processes are tested.

    ii.   Finding errors in the interaction between actors (users or systems).

    iii.   Identifying the maximum and minimum values for every input field.

    iv.   Technical automation and embedded software industry

        a.   i., ii, iii are correct, iv is incorrect.

        b.   I, iii are correct and ii, iv are incorrect.

        c.   i., ii are correct, iii and iv are incorrect.

        d.   i., ii and iv are correct and iii is incorrect.

Q11. Which of the following is not described in component Test?

a.   Equivalence testing

b.   Decision coverage

c.   Stress testing

d.   Syntax testing

Q 12.Which approach should be most effective and efficient in terms of defining test procedures with a highly experienced tester with a business background when a project is under severe time pressure?

a. A high level outline of the test conditions and steps to follow.

b. Every step should be explained in detail.

c. Detailed documentation of the test cases and steps to follow.

d. A high level outline of test condition and details of steps to be taken.

Q.13.which of the conditions is not true?

a. 100% Path coverage will imply 100 % statement coverage.

b.100 % Decision coverage will imply 100 % statement coverage.

c. 100% decision coverage will Imply 100 % Path coverage.

d. 100 % Path coverage will imply 100 % Decision coverage.

Q.14.Consider this pseudo code below. If this was a programming language, how many tests are required to achieve 100% statement coverage?

```
.      If A=10 then
       Display_message A;
       If B=5 then
Display_message B;
       Else

       Display_messageC;
       Else
Display_messageC;
```

a.1

b 2

c 3

d 4

Q15.Code coverage is used as a measure of what?

    a. Test effectiveness

    b. Time spent on testing.

    c. Number of defects

    d. Trend analysis

Q 16. Consider the following

IF my flight is a cheap-day return

Then catch flight after 2:00 PM

Else catch any train

Buy a coffee and enjoy the Journey

    a. SC=1, BC=1

    b. SC=2, BC=1

    c. SC=2,BC=2

    d. SC=1, BC=2

Q 17. Consider this pseudo code

        Read X

        Read Y

        IF x=y Then,

        Print " They are equal"

        Else

        Print "They are different"

        ENDIF

    a. SC=1, BC=1

    b. SC=2, BC=1

c.   SC=2,BC=2

d.   SC=1, BC=2

Q 18. Go to an ATM machine.

If the ATM machine is not working

Then call repair centre to fix

Else insert your ATM card

If PIN is correct

Then select cash withdraw

Else display message " Card is invalid"

a)   CC = 2, SC = 1, BC = 2

b)   CC = 2, SC = 2, BC = 2

c)   CC = 3, SC = 3, BC = 3

d)   CC = 3, SC = 2, BC = 3

Q 19.If your task is to test the status tracking in a defect management system, which keeps track of status of every defect registered and imposes the rules about changing state. which method would be best?

a.   Use case testing.

b.   Logic-based testing

c.   State transition testing.

d.   Acceptance testing

Q 20. What is the reason behind missing bugs even if we achieve complete statement coverage?

a.   If you go through the FALSE branch of the statement as the failure occurs only if you take the TRUE branch of the statement.

b.   The failure depends on the program's inability to handle specific data values, rather than on the program's flow of control.

c.   **Both A and B**

d.   If you do not test the code that customers are unlikely to execute.

Notes:

5. Test management:

Software testing is not single step activity it is a sub process of software development life cycle. But, software testing also involves stages, methods and techniques to be followed. In this chapter we clearly look at how to manage testing in a company with six sections, how to manage a testing team and testing. In the later stages we will focus on test strategies, planning and estimation of testing. The third section states the test progress monitoring, test reporting and test control. The fourth explains configuration management and its dependencies to testing. The fifth explains the topic of risk testing, project and product risk factors and how testing affects and is affected by product and project risks. The sixth and the final section deals with the defect reporting and defects tracking in a testing process.

5.1 Organizing the Testing Team in-house v/s outsource testing:

In this section we will look at different type of testers and testing team can be involved in a project. Roles and responsibilities of each member in a testing team, we will also explain the different designations in a testing team like Test coordinator, Test lead, Test manager and a software tester with their roles and responsibilities. In this section we will firstly look in to how testing team can be organized in a project and we will look at benefits of independent testing over the integrated testing team.

Different organizations follow different approaches to organize their testing teams in some companies we might see the programmers who test their own code, and in contrary some companies may have separate testing team but still working with the developers and reporting their defects to the development lead. We may find a independent in house testing team who works independently without associating to the developers and reporting their defects to the portfolio manager or project manager, looking at the complete independence we may find a separate testing team within or outside the company equally valued along with development team where we can find specialist testers like Automation testers, performance testers, security testers and ETL testers....etc. these separate outsourced testing team may be located within the company or might be outsourced to third party testing services companies who will work with contract.

Now, we will discuss the benefits of having an independent testing team rather than programmers who test their own code and risks involved with independent and integrated testing teams as well. We will also discuss the benefits of in-house testing over outsourced testing teams in a company's testing process.

Independent testers who are software testers by profession can look at the application in a different perspective; they will understand the user requirements clearly and will use the software like a basic user to explore the hidden bugs. Testers can also think with out of box attitude to break the application in a negative testing approach where the programmers cannot. Testers will know the priority and critical bugs, generic bugs that may occur in mere future rather than a developer. More over developers who act as tester will have the parental feelings and will think like Build success test passed, I am confident that my code will work without any errors it doesn't showed any errors during the execution. Programmers can't the complex problems like testing again and again in a different approach for hidden bugs. Programmers will only in to positive ends of the project they are not capable for negative testing like software testers. So, it is always better to have a separate testing team rather than developers acting as testers.

In Team level, independent tester who reports his bugs to the project manager may look for good name and fame, appraisals from higher officials compared to test lead or tester in a development team.

There are also some risks associated with the independent testing team compared to integrated testing team within the development department. Those risks are :

Communication problems may arise due to high focus on bugs and reporting rejected again and again may get an irritated feeling from the programmer's .so, tester should always be very clear and precise with his bugs when communicating to programmers. A detailed list of steps to reproduce the error, screenshots if necessary... Vague descriptions of errors that cannot be reproduced or have unclear steps to reproduce will very quickly spoil the developer-tester relationship.

Sometimes, programmers may also ignore their own responsibility towards the quality of product and will say we have a separate testing team now, so why I have to do unit and integration testing?

These kinds of problems when there is an incapable test manager who is inefficient to manage the priorities of bugs. These above scenarios also suggest that for a better communication between testers and programmers a new Defect tracking team should be there in a company which comprises of both testing and development team leads to evaluate the correctness of bugs before reaching to developer and in defect reports using the tester name and developer names should also be replaced with ID numbers for a better relation between programmers and testing teams.

Sometimes, in some companies testing will be outsourced to specialist third party testing companies which is very useful for saving hidden costs like recruiting, training and tools costs etc. we will also have great flexibility to increase or decrease the size of testing team depending on the client requirements and delivery date which really speed up the testing process with outsourced testing services. Usually outsourced testing services offer certified and experienced testers who can do wow in testing compared to in house testing. There are also some risks associated with outsourced testing like no guarantee to implement tests and for test coverage and for future bugs. It is always better to choose near shore testing company to carry out our testing rather than choosing a company in other country.

There is no specific standard yet to organize the testing team in a company. For every project we have to assess whether to use in house testing team or to use outsourced company testing services? whether independent testing team or integrated testing team? By considering various factors like risks associated, budget allocated for testing and level of testing needed etc.

5.1.1 Role of Test lead:

We have discussed the ways and approaches to organize the testing team in a project and their benefits and risks. Now, we will look at roles and responsibilities of test lead in a project.

- Involving in planning, monitoring & control of test activities.

- Responsible for preparing test plans, test strategies and test objectives.

- Estimate the amount of testing to be done and time needed for testing.

- Scheduling the tasks between testers and assigning roles within the testing team.

- Selection of testing techniques and methods to be followed in testing process.

- Deciding the need of automation tools and selection of Automation tools?

- Deciding when to automate? And when not to automate?

- Providing training and recruiting additional staff if needed in testing department.

- Negotiating with company management for any additional resources.

- Monitors testing activities like Test planning, Test design, execution and Test reporting.

- Arranging the testing environment before test execution.

- Reviewing the tester's test case and defect reporting, time log and status documents.

- Reporting the status of testing to stakeholders ex: project manager, clients etc.

These are the main responsibilities for any test lead associated in the project. In some companies test leads are also called as test managers or test coordinators. Test lead should be approachable always with good communication to project leads, clients and development leads.

5.1.2 Role of a software tester:

- Analyzing the software requirement documents and getting clarified from business and system analysts on any doubts.

- Become familiar with the software under test

- Understanding test plan and test strategy documents.

- Communicating with test lead for better understanding on his roles and responsibilities.

- Arranging the test bed with required software's and hardware's needed for testing.

- Designing the test cases and scenarios

- Executing the test cases and scenarios and documenting the defects in reports.

- Performing re, regression testing to the modified builds.

- Maintain the test log documents.

- Updating the test cases and scenarios based on the changed requirements or discovered defects.

- Reporting the work progress and any problems to the Test lead.

- Prioritizing the defects in defect reports.

- Thinking out of the box in testing and analyzing the software with user perspective.

- Keeping him/her updated with latest testing technologies and tools and approaches to be used in software testing.

5.2 Test planning, Estimation &Test strategies:

In this section, we will look in to the areas of Test planning, levels of Test planning and its objectives. We will clearly discuss the objectives of Test Plans and its contents. We will present a standard template for Test plan. We will also look in to the areas of different test estimations and test strategies and exit criteria's for testing and different aspects in the process of Test planning.

5.2.1 Purpose and Importance of a Test Plan:

First we will look at, what is a Test Plan? In simple words Test plan is a plan for testing process to be carried out. For example if we want to build a house then we will go the architect for a (blue print) plan to build. In the same way here Test Plan manages the whole testing process. Test plan describes the activities and schedule the activities to be followed and methods to be implemented in the testing process. Usually Test plan do not contains test cases, test scenarios or test designs it only contains the ways and approaches to be followed and scheduling etc stuff. Test plan has different definitions across the globe in simple terms it is a future plan of testing activities to be carried out. Test plan is also known as Test protocol in some companies. Test manager or Test Leads are responsible for preparing Test plans before the Test initiation.

Why Test Plan is important? Test plan describes entire scope of testing so, the success of testing in a project really depends on the effectiveness of test plan. Test plans are written mostly for:

Preparation: laying out the ground work for the testing process, assuring that all the resources are there for testing process like manpower, tools, hardware and software requirements. This process helps us to think on the critical aspects of this testing process like scheduling, delivery time areas to be tested effectively etc.

Better Communication: These Test plans itself acts as a communication bridges between stakeholders and testing team. Since, they will let us know about the tools that testing team were using and approaches, methodologies , techniques following by the testing team in this process. We will distribute one or two drafts of test plans to the development leads in reviews meetings so, that they can have a look and clarify their conflicts regarding the Testing schedules, tools and techniques etc.

Quick Adaptability: A clear written Test plan helps us to adapt the changes quickly. In the early stages of project we gathered some requirements for our testing especially in testing methodologies like Agile changes occurs frequently, so with a proper written Test plan it is easier to update the changes by reviewing and updating the test plans according to the new requirements.

Reduce complexity: For vey complex projects which will run for many years it is very hard to carry out the testing process without any documentation. Test plans reduces the complexity of very large by avoiding replications in work and setting up clear expectations and objectives in the early phases of testing like What is in scope? What is out of scope? Objectives and deliverables from testing.

Test plan do not have any particular standard many companies uses customized Specific Test plans to suite their own requirements although there are plenty of IEEE standards available. Let us look at this IEEE standard 829 Test plan.

IEEE 829 STANDARD TEST PLAN TEMPLATE

Test plan identifier	Test deliverables
Introduction	Test tasks
Test items	Environmental needs
Features to be tested	Responsibilities
Features not to be tested	Staffing and training needs
Approach	Schedule
Item pass/fail criteria	Risks and contingencies
Suspension and resumption criteria	Approvals

Figure: IEEE 829 Standard Test Plan template

Test Plan identifier: which provides a unique identification to the test plan which can be given with project name or using numeric's etc. Ex: GetSkills_TP1.

Introduction: A brief introduction about the software and users of the software and can include references for any related documents that can support this testing process like project plan, design documents, requirements documents etc and defining the complex terminologies used in those documents.

Test Items (Functions): These are the things intended to be tested within the scope of a test plan which also includes a comprehensive list of items to be tested in the project and configuration management can be also included with software version numbers the software support etc.

Features to be tested: This is a list of features, characteristics to be tested in AUT with user perspective like login feature in various conditions and money transfer functionality in a secured manner in a banking application, performance, security, compatibility etc.

Features not to be tested: List of characteristics, features that can be omitted without testing. Identify why the features are not needed to be tested and provide the reasons in detail like Low Risk, Not included in release of software, incapability to test needs a special tool.

Approach: This is the overall strategy of this test plan with overall rules and processes should be identified, items and their features to be adequately tested. Mention the overall approach of this testing. Specify testing types, methods and methods like black box, white box, manual or Automated etc.

Item Pass/fail criteria: specify the criteria to be used to determine whether each test item has passed or failed during the testing and include any screenshots related to the issues with severity and priority status.

Suspension and resumption Criteria:

This is also known as entry and exit criteria for a project which describes the criteria to be used to start the testing process and how to know when to stop testing process. Usually exit criteria have some requirements which will be discussed in further sections.

Test Deliverables:

This explains about the list of documents to be delivered from the testers like test cases, Test scenarios and test log files. It also explains list of documents, charts, screenshots to be presented to the stakeholders from the testing team on a regular basis during testing and when testing has completed.

Test tasks:

Describes the set of tasks to be followed within the testing team from test lead to tester in the process of testing.

Environmental needs:

This explain the environment required for test execution like any special hardware or specific versions of software requirements like Simulators, static generators, Mobile phones, servers, databases, automation tools etc will be specified.

Responsibilities:

Discussing about the roles and responsibilities of each member within testing team like as follows: Who is in charge? and who should report to test lead, Who should be asked if any requirements are there?, who provides required training?, who is responsible for the risks?, who makes decisions like critical bug/non critical bugs in the team.

Staffing and Training Needs:

Identify the training needs to the testing department for any new software's or Automation tools to be used. Conducting training sessions if required. Specify the staff needs and assessing the current department staff and recruiting new staff if needed to deliver the project within the time limit.

Risks and contingences:

 Identify all the risks within the project and specifying the alternate risk reducing plans for each risk specified. For example: requirements analysis will complete by next month 21-xx-xxx, and, if the requirements change after that date, then these following alternate actions will be performed to reduce the risks like resources will be added to testing team or updating and reviewing test cases again according to the new requirements etc are the alternate plans for depletion of each risks specified.

Schedule: scheduling the tasks within the team, Goal setup and time setup for every test cycle and assigning the tasks to the testers within the team and scheduling all the phases of testing activities with a specific time.

Approvals: Specifying the roles of persons responsible for approvals like who should approve the test plan and who should approve the testing process as complete and approve the project to the next stage?

So, far we have seen how the test plan will be written and the importance of test plan, a standard template of a test plan with its contents. Now we will look at test estimations concept.

5.2.2 Test costs Estimation:

Testing is not a separate process it involves within the development life cycle and testing is a sub process in a project. But, Software testing involves different phases and stages which each phase constitutes a different process starting from requirement analysis, test planning, test designing and test execution, test reporting and test closure. Each of these phases again may involve sub process and sub tasks in it. Let us first look at what testing process includes, and how test life cycle will be carried out to determine and estimate the costs involved in each phase.

Software Testing is not a single step activity it contains levels and phases to be followed. For every development there must be a testing activity too. Likewise developers follow SDLC models testers also follows STLC process starting from the requirement analysis phase to test closure phase. But, there no standard STLC model to be followed every company uses their own customized model to fit their needs.

Software test life cycle model may be used at any stage of testing for any test activities if we have seen a new risk and some changes have done or some of the requirements are changed again we follow the STLC model by gathering updated requirements from requirements phase and updating the test plans in test planning stage, test designing stage involves designing the test cases and test scenarios for the updated modules these phase will be followed by test execution where testers will execute all the designed test cases and will look for the errors. If any errors found will pass through the test reporting phase again testing the fixes and finally this STLC ends with test closure activity when all the major issues are resolved and all functionalities are tested then the process testing will be ended with a Test closure note.

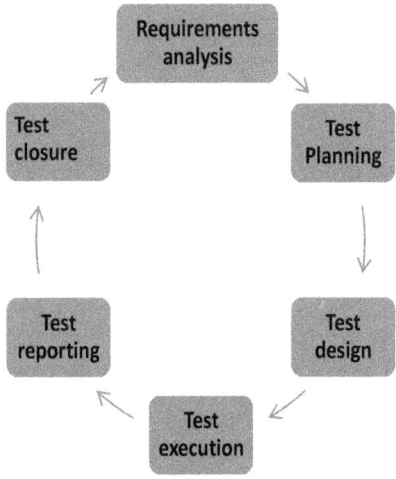

Figure: Software Test Life cycle

So, now we have seen the process involved in the software testing neither it is not a single step activity to be performed nor it is not a separate process apart from the project. Test estimation activity can be carried out in a bidirectional traceability approach to estimate either the testing effort or testing costs. Bidirectional traceability is implemented from high level to low level and from low level to high level. i.e. from requirements to end products and from end products to requirements. When requirements are managed well, traceability can be implemented from the source requirement to its low level requirements and from lower level to the source again.

Let us look at a real time scenario in our project test manager identified that most critical area to be concentrated for the AUT is performance testing or security testing. Now in the execution phase you decided to conduct performance testing as well as security testing. Then again staring from the requirements stage your Test estimation for performance and security testing process should be carried out with considerations like How long will these security testing process takes, do we need to recruit any specialist testers who able to do performance testing, tools needed for performance testing and their cost, training costs to the staff with respective of new tools acquired, Time taken for writing automation test scripts and execution and also identifying the hidden costs like recruiting costs, test environment set up costs, travelling costs if needed, and risks involved in updating the new requirements and test plans, time required for re and regression testing phases etc. so, in any project usually divide the total project in too many sub tasks while giving the estimation time or cost. If we divide and look at each specific task then it will be easier to provide estimates precisely. say for example if you require 20 days to complete one test cycle execution then keep some buffer time for unexpected incidents and give the estimate as 25 days to avoid last stage risks in any project before the delivery time.

5.2.3 Techniques for estimation:

We have seen how test effort and cost estimations will be involved and in a project planning. Now, we will focus on the different techniques used for Test Estimation.

There are no specific standard of estimation techniques globally every company follows their own customized strategies in estimation process. According to the ISTQB syllabus there are two Test Estimation techniques. They are consulting the experts or Analysts who are well versed with the project metrics and estimations and other technique is estimating the costs based on the previous projects or comparing with the market prices and other company prices. Let us look at each of these techniques in a detailed manner.

Experts from all the departments are gathered like designers, developers, project managers, business analysts and testers are gathered and will discuss the complexity of the project and measure the size of software application based on the number and complexity of user estimated and tested, delivered modules etc, it can be also judged with staff metrics like total no of testers involved in the project, time taken for testing, Total no of cases written and executed , specialists recruited and their costs, tools purchased and time taken for automation scripts execution and test environment setup costs, no of total tests done and no of test cases executed per day and total no of days the entire testing process has taken, developers and testers ratio involved in this project and total time duration etc will be considered in estimating the costs this technique is known as expert decision or Bottom up approach in estimation.

The other technique to estimate the cost and effort for your test project cost and estimation is experience analysis. Estimate the effort and cost from the previous projects your company did in the same way. Analyze the time and cost that you have charged in previous similar to assess the current project estimation. Gathering the industry data as well analyzing the market rates will also help for the precise estimation testing effort in a project.

Test estimation plays a key role in the process of testing as well as in building good relation with the clients and also will have an impact on the entire test life cycle for the project involved. So, maintain the precise metrics like how many hours worked, how many testers involved and how many test cases written, executed, passed and failed. Communicate with people within the team and with developers as well as project managers before estimations. Usually test managers use tools like Microsoft project and word pad, sticky notes tool to note down the metrics of estimation. There are some automated cost estimated tools that can be also used for precise estimation of project. The new trend in test effort and cost estimation is usage of Mind map tools.

5.2.4 Factors affecting test effort:

Testing is a critical and complex job in the project which has several factors influencing the process of testing. In this section let us look at various factors that can affect the test effort. Test strategy or test methodology implementing in the project will show a great impact on the test effort.

Moreover these test estimations and Test planning's are dynamic and may changes at any stage of the project like in the initiation or at the ending stage when there will be no time of delivery. So, test manager should have a clear future view of the project and should be ready to take up challenges to suit their testing approaches according to the changing dynamic conditions. Sometimes expert project with no testing knowledge and experience may also fail in estimation process of allocating resources and time for testing. It is always good to combine the test estimation techniques with better understanding of factors that can influence the project in future like availability of resources, time, dependencies, effort etc. some of these factors can suddenly slow down the process of testing effort. The various other factors that can influence test effort are as follows:

- Improper documentation of requirements and specifications documents like SRS and BRS may increase the complexity of testing team. So, that Tester may require some additional effort and time for better understanding of the project behavior and functionalities before signing in to the testing phase.

- The chosen development and maintenance life cycles like waterfall model, agile model will also influence the Testing effort.

- Commitment of tester's in a team will fast up the process if the test deliverables are delivered within the time limit in every phase will fast up the process.

- Commitment of developer's team to respond for the bugs reported by the tester's team and their fixing time period (Bug Life cycle) should also be considered for test effort estimation.

- Critical testing techniques like performance testing and data volume testing and automation testing are very complex and require additional time to perform.

- Geographical location of the teams is also an important factor to be considered especially if the testing teams are working from different locations in different time zones, management of testing work will also be considered as an important factor when some of the testing work like Automation, security testing is outsourced to other Third party testing services.

- Skills of the professionals involved in the project also differs the time factor of the projects development as well as testing phases.

- Proper execution earlier phases in testing like unit testing and integration testing will make the process easier.

- Identification of reusable test data and documents from the similar kind of previous project will reduce some work burdens.

It is important for the test team or test manager to consider how each of the above factors will affect the estimate before preparing the Test effort estimation. Experience is always a key factor to best test effort estimation. But sometimes test efforts and estimations will go wrong. So the test manager should clearly monitor the estimations and deliverables in every phase of testing in the entire project.

5.2.5 Test strategy or approach:

The selection of Test strategy plays a key role in influencing the correctness and quality of the entire project. Test strategy explains about the things like what we want to achieve with this test strategy in this project and how we are going to achieve it?

Test strategy explains the test levels to be performed and the test activities to be performed within the project to ensure quality. The purpose of test strategy is to explain the critical things and challenges in the testing process to the stakeholders before initiation. Test strategy can also be called as Test approach and Test architecture and Test handbook sometimes. There are several test strategies available and have been practicing in the real world. But sometimes the entire test process may fail if the correct test strategy is not adopted for the project. Let us see the different test strategies available.

Analytical test strategy:

Analytical test strategy involves the two main things called risk based testing or requirements based testing which involves analysis before delivering the project to the customers. this test approach is analysis based on any strong factor that can influence the testing process. For example if requirements are influencing the project then clear analysis id performed on requirements to identify the most critical requirements to be tested first and least critical requirements can be tested later. In other way, analyzing the risks involved in the project and executing the most critical risks first in a top down approach. This strategy can be considered as a proactive process where we are assessing the risks and requirements in the early stages to prevent further failures.

Model Based strategy:

Here tests are performed based on either mathematical or statistical model of objects functionality. Develop a model with specific inputs and anticipated results and if our model performs according to the specified anticipated conditions in the model then our predicted model is our working fine. There are several formal and informal methods to be carried out in this strategy. Model based strategy can also be considered as a preventive strategy.

Methodical strategy:

This strategy depends totally on the predefined method to be carried out. The predefined method can vary widely or may present with checklists of executions like error guessing and experience based approaches. The entire tests will be planned, designed, executed and reported according to the method specified in the beginning. Actual testing effort may start involved from the beginning or later stages of SDLC and test implementations.

Process or standard compliant method:

As the name suggests this is a process oriented test strategy technique where for example if your company is following the standard IEEE 829 methodology for testing process as an alternate to fill the gaps you may allowed to use another methodologies like Agile or another in between. Again at the actual test effort started before or after the SDLC model implementation after choosing a specific methodology. Hence this strategy is also considered as the preventive one.

Dynamic Strategy:

Dynamic strategy is also known as heuristic strategy which do not involves any specific methodology or rules like process strategy and methodical strategy. For example if we specified a simple guidelines before the process of testing then they might changed based on the requirements may adopt Exploratory testing techniques in the middle of process to concentrate more on the specific critical areas where tests are executed and designed simultaneously. Hence this strategy comes under reactive.

Consultative or directed:

As the name suggests in this strategy we may consult non testing groups like developers and business analysts, domain experts, consultants to know the Testing process and testing methodologies and effort involved in testing process. Here the entire testing will be directed outside of the group like developers head, experts or consultants like security specialists etc.

Regression Averse:

This strategy recommends designing the testing such that all the regression stage bugs should be detached in the earlier stages. This type of test strategy will require Automation tools and Automation testing involvement in the process.

All of the above discussed strategies some of them reactive and some of them are preventive strategies depending on the circumstances and conditions, risks involved in the project we will choose the specific Test Strategy.

For example consider the risks involved in the project and level of risks. For a developed project regression testing phase is important to find out the new the bugs occurred in modifications. For a new project analyzing the risks properly may give new bugs. In such cases choosing the analytical method or regressive averse methods will make sense.

When there no skills and lack of proficiency is there in your team members as well there is lack of time in such cases compliant test strategy will work better.

Dynamic test strategy can also be choosed if there is lack of time and if the main objective of the stakeholders is to find as many as bugs within the specified time limit then behalf of sticking to any certain technique following the dynamic test strategy will give you the quick results.

In some cases stakeholders and clients will have certain regulations and rules in such cases we have to stick on the entire testing as per their regulations. To be successful in such situations test strategy like Methodical techniques will be used.

We can also use model based test strategy technique if we have any previous developed similar legacy system that can acts as a model for the test process.

These are the rules to be considered before choosing a specific test strategy technique to achieve better results.

5.3 Test progress monitoring and Test control:

In this section we will focus on how to monitor the progress in testing process for a project and the metrics associated in Test monitoring and Test control and Test reporting. We will deal with the techniques associated with test progress monitoring and test control techniques. We will also present test reporting process and discuss about defect report templates and its metrics.

5.3.1 Monitoring the progress of Testing process for a project:

After developing the test plans and test activities associated with it the testing process should continuously monitored to know about what actually happening is? this is called Test progress monitoring. This will also help to know whether our testing process is happening within the specified test plan or it is going beyond that. The purpose of Test progress monitoring is provide the actual better feedback to test managers, Test leads, Project managers and clients about what actually happening in the testing process, it also helps to have a clear visible view on the entire testing process.

Test progress monitoring can be done either manually with the help of spreadsheets and word documents or with the help of specialized automation Test management tools. Manually by assessing things like how many test cases are written and executed by the testers today and by using metrics like code coverage and exit criteria will help to monitor Test progress within the project. The test progress monitoring will also help to find the exit criteria of the project like 50% of requirements given are tested and there is still 50% to be tested. Ultimately the final goal of test progress monitoring is used to track the progress of testing and to assess exit criteria.

The other way to monitor the progress in testing is use of IEEE 829 standard Test Log document. Let us look at the standard template of Test log which explains the daily/weekly work status of a tester.

IEEE 829 STANDARD: TEST LOG TEMPLATE

Test log identifier
Description (items being tested,
    environment in which the testing is
    conducted)

Activity and event entries (execution
    description, procedure results,
    environmental information,
    anomalous events, incident report
    identifiers)

Figure : Test Log Template(Dorothy Graham)

Test Log Identifier

· Unique "short" name or ID number for the log with version and date

Example:Test_login_TL1_23-june-2014

Description

· Items being tested with supporting reference materials like Test Case Document

· Executed by Tester, observer with Date/Time

· Tester

Activity and Event Entries

· Beginning of each significant activity with Date/Time

· End of each activity

· Execution description

· Procedure executed (reference to its location)

· Tester, developer, observer

· For each execution, log all relevant information

· Error messages, aborts, interventions with screen shots if possible

· Location of outputs

· Result status (success, failure, unknown)

· Environmental information

· Any changes or substitutions from requested environment

These are the contents that each test log developed should be maintained, there is no specific Test log document many companies uses their own customized documents and usually at the end of the day the testers in testing team individually will forward their Test log sheets to the correspondent authority. In some companies they will report test logs weekly or in some these test log documents may be forwarded and reviewed for the every module tested by the tester. The corresponding support documents like Test case report developed by every individual tester will also be attached along with the test log document while forwarding to the responsible authorities. let us look at the standard Test case template.

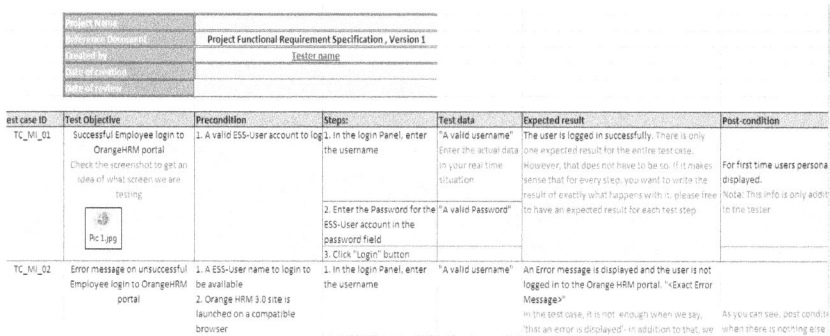

Figure : Test Case Template

The standard test case will also be attached along with the Test log document for future references and Test case document may also include the bugs with specific Bug id or Bug report can also be maintained in a separate report with status and severity and priority details.

Some of the specialized automation test management tools will clearly show the Test progress with the help of graphs and pictures with How many passed/ how many failed cases and their percentages as well. Usually large projects may require Test management tools to provide in detailed Test logs or Test progress reports.

Test progress monitoring cannot be achieved only with the help of Test log document itself there are also several supported documents like Bug/defect/incident report document and test cases document to assess the projects current situation. We can also evaluate the progress of testing by following defect metrics like How many Bugs identified and reported?, How many bugs are fixed?, still how many bugs are yet to be fixed from the developers?, How many modified modules are re-tested? These are some of the guidelines to be used while assessing the test progress within a project. All the supported documents are thoroughly examined and evaluated and also by considering the risks associated with project, budget, code coverage and the extent of testing done and the extent to be done are the factors that can play key role in Test progress monitoring.

5.3.2 Reporting the Test status:

Test monitoring process will ensure to know the current progress of testing in a project and helpful for estimations like exit criteria but, Test reporting is the process where every individual tester will report test metrics in a detailed format to update the reader, customers regarding the testing tasks undertaken.

- What has happened during a given period of time, e.g. a week, a test level or the whole test endeavor, or when exit criteria have been met.
- Analyzed information and metrics required to support recommendations and decisions about future actions, such as:

1. An assessment of defects remaining and defects solved

2. The economic benefit of continued testing, e.g. additional tests are exponentially more expensive than the benefit of running

3. Risks involved, risks analyzed and future risks

4. The level of confidence in tested software, e.g. defects planned vs. actual defects found. The IEEE 829 standard includes an outline of a test summary report that could be used for test reporting and we can also use the above specified documents like Test cases document or test log documents as well.

The information gathered can also be used to help with any process improvement opportunities and for taking decisions with stakeholder involvement. For example the critical bugs can be identified with Automation Regression testing but requires time and additional resources, In such cases with Test report we can ask for the real customer decision whether to continue Automation or to terminate and deliver the project as soon as possible. This information could be used to assess whether:

- The goals for testing were correctly set (where they achievable; if not why not?)

- The test approach or strategy was efficient (e.g. does it ensure there was enough coverage?)
- The testing was effective in ensuring that the objectives of testing were met.

IEEE have standard Test report template but companies will follow their own methods to write test reports either with visuals like Graphs, bar charts to show the percentages as well or they may use standard excel sheets or word documents too.

Let us have a look at the Standard IEEE 829 format of Test Summary report.

IEEE 829 STANDARD:
TEST SUMMARY REPORT TEMPLATE

Test summary report identifier   Evaluation

Summary   Summary of activities

Variances   Approvals

Comprehensive assessment

Summary of results

Figure : Test summary report (Dorothy Graham)

Test summary report Identifier:

A unique number for the summary report for identification, with stuff like versions and build numbers, Date/Time etc.

Summary:

An overview and summary of the project and activities have done so for.

Variances:

Recording any variances from previously agreed functionalities, especially in the areas that may cause concern for the clients in accepting the product and test results. If any variances occur include the supported documents that can explain the reasons for the deviation.

Comprehensive Assessment:

Provides a brief assessment of the Application under testing (AUT) name, version no:, Build no:, and may also include Iteration Id etc in a comprehensive manner.

Summary of Results:

Try to summarize the overall results of the testing process. Identify all the errors or incidents found during testing with their Specific ID numbers, priorities and statuses, discuss the error fixes, and any unresolved incidents.

Evaluation:

Provide an overall evaluation of the project based upon the test results and the number of incidents resolved or unresolved. Summarize the testing activities and the next stage of testing steps.

Summary of Activities:

Provide only the major testing activities of the project and bugs, resolutions.

Approvals:

Discuss about the approvals of testing activities to further level in the project and also discuss about Staff, Tools and Time approvals with their role and designation.

These are the contents that can be included in IEEE 829 standard Test summary document along with this document tester's team can also include any supported documents like Bug reports, test cases document and test plan documents as well for the better understanding of stakeholders.

5.3.3 Test control:

Test control is the test management tasks required throughout the testing process in order to keep the entire testing to match with the software development process, customer requirements, budget, the needs of the project, and the needs of the company. Test control is guiding the actions with a motive of achieving the best in any situations.

Keeping the entire testing team and the testing process to meet any kind of unexpected things that may occur in the project is Test control. Because everyone will plan something that may or may not occurs so keeping ourselves and staff to adapt the new challenges and changes is very important for the success of project. Before starting the project we may write perfect Test plans and Test strategies but they can be changed or updated according to the situations. In such situations our team has to adapt the new conditions quickly without inferring the delivery date and quality of the project.

Some examples like ever changing requirements even conduct the regular requirements analysis sometimes the requirements may change or updated with some new functionality where the tester's team should be aware again to test the updated requirements within the time. In some situations like in the phase of Test delivery and Test execution some of the key person like test lead is changed in such cases it's very hard to find the new test lead as soon as possible and to adapt the current project conditions. So, Test manager should be aware of all such risks that can happen in any way at any time during the project.

In some cases where delivery time is getting closer, developers may react slowly to the bugs reported by the testing team and the fixes may be delayed also. Even though the Testing team has worked up to their caliber some unwanted delays may be created which should also be considered in test control.

Some of the Test Control activities include:

- Making the decisions according the conditions(new changes)

- Prioritize the Testing activities when there is limited delivery time left

- Change the schedules of testing based on the staff capabilities, availabilities (If any one of the staff changes the company in last phases).

The worldwide leader in tester Rex Black specifies "Think of the test plan as a roadmap, with the starting location and the final destination clearly indicated. This roadmap will help you drive to your chosen destination. However, throughout your drive, you should plan to stop at traffic lights, mind your lane and speed, adapt to unexpected events (such as pedestrians stepping into a crosswalk), and even adaptively overcome errors in the roadmap. A good Test plan makes the Testing easier". (Black, 2011)

5.4 configuration management:

In this section we will look at Software configuration management (SCP) how all the software's and hardware's within the team from both developers and testers are aligned for better results with their different versions and configurations. **Configuration management** is the detailed recording and updating of information that describes an enterprise's computer systems and networks, including all hardware and software components. Such information typically includes the versions and updates that have been applied to installed software packages and the locations and network addresses of hardware devices. Special configuration management software is available. When a system needs hardware or software upgrade, a computer technician can accesses the configuration management program and database to see what is currently installed. The technician can then make a more informed decision about the upgrade needed.

Why Configuration Management?

Why do we need configuration management really? Let us think about it, software testing is detecting the bugs from the software before delivery for a better quality. But the detected bugs should have certainty and should be reproducible. Software tester might be aware of this scenario where testers will report the bugs and developers will not accept because the reported bugs are not reproduced in developer's environment. This happened because software tester may use different software and hardware with various configurations and version numbers and got a bug but, there is no certain rule that the developer will also use the same configuration system to check the reproducing capability. So these types of issues may arise in workplaces if software and hardware systems are not managed properly.

The ability to reproduce a bug requires the same conditions, that are things such as a matching environment, the same software, and the same conditions before running the test and the same data.

Software test configuration also manages things like Build Numbers and version numbers especially when testing a large project it is very important to maintain the Build ID's for every new build released from the developers with fixes and also in bug reporting tester's should be clear about the bug in which conditions the bug is identified along with the details like Software and Hardware installed in their systems with their configurations and version numbers along with the browser version number for reporting web application errors. These all the above specified things can help to manage the hardware and software configurations in a project. IEEE 829 specifies standard Item transmittal report to be used for better configuration management.

IEEE 829 STANDARD: TEST
ITEM TRANSMITTAL REPORT TEMPLATE

Transmittal report identifier
Transmitted items
Location
Status
Approvals

Figure : ITEM TRANSMITTAL REPORT (DOROTHY GRAHAM)

IEEE 829 Standard specifies an Item transmittal report for the configuration management process which includes

5.5 Risk and Testing:

In this section we will focus on risks involved in the project that may show great impact on the project, the risks that might occur and how to determine the risks involved and risks may occur. We will also look at various risk factors and levels of risks, project risks and product risks. This section also focuses on risk analysis and risk management.

5.5.1 Risks and levels of risks:

We have been using the word risk several times, what is risk? Risk is a chance of event that can occur in future with possible potential outcomes. We can't predict the future of the project but we know the probabilities of uncertain things. Risk is an unwanted damage that can occur at any time in anyway of the project like damage to the corporate infrastructure, loss of faith from the customers, uncovered testing defects which may cause serious failures in future, delay in delivery time etc. Risk tells us the probability of failures that may arise and damages in terms of faith or money, company name and fame. In order to prioritize testing we need to understand the risks and apply strategies to overcome them intelligently.

Every risk is associated with some negative outcomes, we need to consider those negative outcomes in risk analysis. The impact of the risks depends on the how serious and how critical the risk is. suppose in a project testing bugs are detected quite usually and will be fixed before delivering the project to the customer. But after the project delivered if any uncovered bugs that can cause a great damage to the customer will definitely show adverse impacts on the company responsible for testing the project. So, in some stages occurrence of risks are quite common but the level of risks may be not critical but in the case of failures occurred after project delivery the overall impact and consequences of the risks are very high.

These risks will have different consequences and will show different impacts depending on the stage they might occur. Risks in terms of IT industry can be classified in to two types. They are product risks and project risks. Let us see what the product risks are first.

5.5.2 Product risks:

Product risks will be associated with the software or the system and the possibility of the system or software may fail to satisfy the customer expectations is called product risk. There are several factors can be considered here in product risks like certain functionality specified by the user may not present or may not work properly in the software, unsatisfactory software which have bugs will also cause great damage like loss of name and fame to the company and low quality software will also have several problems like reliability, usability, performance, adaptability and the better security will also be considered as product risks.

Risk based testing is one of the approach which can reduce the possibility of product risks. Risk based is prioritizing the testing activities within the project based on several factors like time, budget, most critical modules, Customer revenue generated modules in a software and also the most errors identified and fixed modules. Based on these factors we will prioritize the testing tasks to be performed when there is no sufficient time and budget for testing. Let us say in a shopping web application Billing and payments are the most critical modules which generate the revenue and billing should be done according to the shopping done. So, these two specific modules are tested with extra effort and other modules with priority wise. Primarily, we identify the risks involved in the project; we will also analyze the risks associated with the potential cost of the projects. So, Risk Analysis is also involved in the process of risk based testing. Risk based starts in the early phases of Software testing because fixing cost for the early bugs is lower than fixing costs of bugs in later stages of a project.

Risk based testing starts with product risk analysis. There are several methods used for this risk analysis are:

- Clear understanding of software requirements specification, Business specifications through studying SRS and BRS documents, design documents and other documents.
- Reviews with the project stakeholders.

Risk-based testing is the process to understand testing efforts in a way that reduces the remaining level of product risk when the system is developed,

- Risk-based testing applied to the project at very initial level, identifies risks of the project that expose the quality of the project, this knowledge guides to testing planning, specification, preparation and execution.
- Risk-based testing includes both mitigation (testing to give chances to decrease the severity of errors, especially for high-impact faults) and contingency (if a risk is happened to be occurred then prepare for alternatives to reduce the impact of risk).
- Risk-based testing also includes measurement process that recognizes how well we are working at finding and removing faults in key areas.
- Risk-based testing also uses risk analysis to recognize proactive chances to take out or avoid defects through non-testing activities and to help us select which test activities to perform.

Risk Analysis can also carried out by clearly analyzing the previous risks and risks that may occur in future by asking the experts like How about the chances of occurring? What will be their impact and what will be this risk severity etc.

For example we are creating a web Application for ticketing system to a Rugby match with the capacity of 500 people can login at the same time for purchasing a ticket to rugby match. But there are risks involved that can be identified in several ways like Kiwis love to play rugby there must definitely more than 500 people who looks for a rugby ticket will use the application at same time, then what will become if more than 1000 people logged in at the same time for purchasing- may be system will be crashed or system will be slow down –if the system is crashed what will the impact, what will the loss if the system becomes slow down what is impact and what are the consequences etc. Likewise Risks can be analyzed and prioritized to carry out risk based testing. Severity of the risks can be marked based on the 1-10 scaling or otherwise using (severe, very high, high, medium, low) methods. It is suggested to note down all the risks in a separate document with their severity and impact level to avoid problems like missing risks in very large projects.

5.5.3 Project risks:

As we are aware that testing is an activity in a project and so it is subject to risk which may endanger the project, so we can say that the risks associated with the testing activity which can endanger the test project cycle is known as project risk.

A popular software tester Montvelisky specified project risks as *"Project Risks* are situations that may or may not happen (risks), if they materialize they usually cause delays in the project's timelines, and the source of these risks may be internal or external" (Montvelisky, 2009).

In order to deal with project risks we need to apply concepts like identifying, prioritizing and managing the project risks.

Some of the risks associated with project are:

1.   Delay in the test build from test team.

2.   Unavailability of test environment for test execution.

3.   Delay in fixing test environment due to lack of system admin.

4.   Delay in fixing defects by development team.

5.   Organizational problems which can be like shortage of staff. Required skills etc.

6.   Major changes in the requirements documents which invalidates the test cases and requires changes in the test cases as well.

For all the project risks the risk mitigation plan should be in place.

For any risk, project risk or product risk we have four typical actions that we can take (Dorothy graham).

▪   **Mitigate:** Take steps in advance to reduce the possibility and severity of the risk.

▪   **Contingency:** Have a plan in place to reduce the possibility of the risk to become an outcome.

▪   **Transfer:** Convince some other member of the team or project stakeholder to reduce the probability or accept the impact of the risk.

▪   **Ignore:** Ignore the risk, which is usually a good option only when there is little that can be done or when the possibility and impact of that risk are low in the project.

Apart from the Risk based testing and project risks and product we should also be aware to manage the risks in a project. We can apply some straight forward common principles here like identifying the risks in the earlier stages of Test planning we can also Standard Test plan template to assess the risks associated and to plan them accordingly. Maintain checklists and prioritize the risks with grades and try to fix those risks by mitigation and contingency, regularly conduct a review meet up with stakeholders to know more about the application being developed.

Most of the tasks related to Risk Management are not complex but they require good understanding of the project and product as well as the strict discipline required to keep following and managing these risks throughout the whole lifecycle of the project.

5.6 Incident Management:

This is one of the important section which deals with Test management activities like How the Bugs are Identified by Tester's?, How they are documented and reported in further stages before fixing the error. We will also discuss how the Standard IEEE Incident report look likes and also discuss the contents of Incident report.

Incidents are also known as errors/issues or Bugs which can be occurred at any time of Testing. This identified error needs further investigation. If the expected output is not equal to the actual output in test execution then it will be considered as an incident and will be reported by the tester in a Bug report or Incident report with the details of the incident.

Let us look at those aspects like how incidents are reported and managed in the process of testing.

5.6.1 Incident Reporting:

Let us know what is an Incident or error or Bug first? Simple Wikipedia" explains an incident as a software Failure/flaw or error that occurs and makes the software to perform correctly and produces an incorrect output. In other ways Software error is violation of customer specifications or requirements that software failed to perform and when its actual output doesn't matches with the expected output is called as a Bug, Error, Defect or Incident.

In the process of software testing, testers will check the Application under testing (AUT), by executing all the test cases written and checks for whether the AUT expected output is equal to the actual output or not if it is not equal then the testers will regard this as an Incident and will report it in to the Incident report. This Incident report will be managed and reviewed by the higher officials in Testing team as well as in Developing team usually called as Defect Tracking Team (DTT) and will assign those incidents to the related developer to fix them if it is a valid incident, otherwise if the incident is not reproducible or invalid then Defect tracking team can assign the status Invalid and closes the incident reported by the tester. It is more common process to find incidents and reporting them in testing but the bug report forwarded by the Tester should be well explained all the intended steps needed for reproducing the incident and severity/priority of the incident, Incident ID, Module name, Incident description and the related documents, screenshots of the incidents and even some times to explain complex incidents screen recording software's such as Active Presenter, Tech smith tools etc can also be used to include video of the incident in incident report.

Incident reports provide the identification steps for the developers to reproduce the incidents which can help for them to fix those particular incidents. A part from this, Incident reports can also be used to check the Test monitoring progress and can be used as documents that can be shown to the stakeholders to explain the testing effort and estimation process. A clear detailed good Incident report helps developer to understand what really the incident and helps for better quality of the project.

```
               IEEE 829 STANDARD:
        TEST INCIDENT REPORT TEMPLATE

Test incident report identifier
Summary
Incident description (inputs, expected
    results, actual results, anomalies,
    date and time, procedure step,
    environment, attempts to repeat,
    testers and observers)
Impact
```

Figure : Incident report template(Dorothy Graham)

IEEE 829 has a standard incident report which can be followed for reporting incidents. Several companies are using sophisticated tools for Incident reporting like Bug Zilla etc and customized Incident reports with screenshots and videos too anyway let us discuss the traditional IEEE 829 Incident report.

Test incident Report Identifier:

A unique number to the incident is given to avoid duplications.

Summary:

A brief summary on the incident is included in terms of how the incident is occurred with which inputs and outputs, steps for reproducing the incident, environment the incident is identified like software versions, browser versions, hardware, OS version etc should also. Tester ID or name can also be attached.

Impact:

The impact of the incident identified in terms of overall quality and its seriousness can be included in the incident report by assigning the severity and priority of the incident. Severity explains the seriousness of the incident where as the priority explains the urgency of incident to be fixed as soon as possible within the developer.

The other terms that can also be included in the Incident report are:

- Status of the incident can also be included which contains (open ,closed, deferred, Re-open etc)
- Attaching the related support documents like Screenshots of incidents, videos.
- Degree of impact on stakeholders especially on customers.
- Date/time and software life cycle in which phase the incident is identified.

The above constraints in a detailed manner can add a great value to the incident report for good understanding by the developers which can increase the quality of software after fixes.

Example Incident Report:

	A	B	C	D	E	F	G	H	I
1	BUG REPORT								
2									
3	Project Name:			Module Name					
4	Reported by			Reported To					
5	Resolved by			Resolved on					
6									
7	Defect #	Reported Date	Status	Priority/Severity	Reference/URL	Defect Description	Method of Operation	Defect Category	Remarks
8									
9									
10									

Figure : Example of Incident report (source:c-sharpcorner.com)

5.6.2 Incident report life cycle:

Every incident reported will go through a lifecycle from the phases from identification of incident to solution of incident. The starting phase of incident life cycle is the NEW phase and ends with closed phase. Let us check what will happen from the starting phase to the incident reported by the tester till the closed phase.

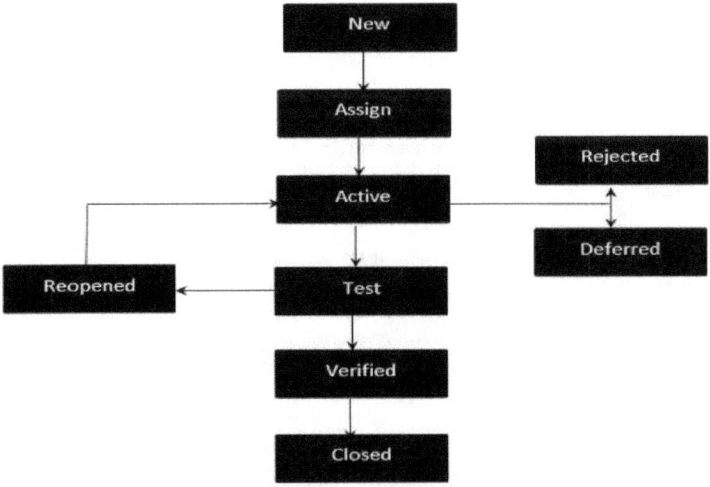

Figure: Incident Life cycle (source: Tutorials point)

As shown in the above the incident life cycle starts with the new phase where any tester who identifies the incident will report the incident to the defect tracking team with incident report discussed in previous section. The detailed content of each stage in the incident life cycle is:

New state:

The initial phase in incident life cycle occurs when the tester reports the incident and yet to be verified by the defect tracking team it is considered as in New state.

Assign:

When the tester reported incident is assigned to the developer team or specific developer to resolve the incident then it is known as Assign state.

Active/Open:

Active state is also known as open state where the assigned developer in the previous assign phase will address the incident reported by the tester and accepted the incident for fix. The tester reported incident is accepted by the developer team as a valid incident then it is considered as Open or Active state.

There are two possible outcomes in this stage those are Rejected and deferred. When the incident reported by the tester is not reproducible or invalid or duplicated then the developer team will reject the incident with status Rejected. In other case the reported incident by the tester is a valid one and also accepted by the developer team but it is not needed to be fixed immediately or incident might have very low impact then the fix is postponed with a status called Deferred.

If the incident in this not deferred or rejected then obviously developer team will accept the incident for fix and will release a new build with incidents fixed.

Test/verification:

After the developer team has fixed the incidents/bugs then again then new build is ready for Re-testing by the testing team to conform whether the incident is fixed or not in verification state. Here the testers will also check for impacts of fixes over the related modules in the software.

Closed:

Closed is the final state of incident life cycle after the verification on new fixes by the testing team if the incident is resolved then the Tester will assign closed state and continue testing and reporting of incidents in other modules of software.

Re-open:

Re-open is the stated where the assigned incident to the developers is not fixed. In verification state tester will check whether the reported incident is fixed or not? If the incident not fixed then tester will again assign the Re-open status to the incident which needs a fix again from the developer team because the incident is not resolved.

Likewise every incident will follow the incident life cycle when the tester reports the incident. This whole life cycle will be monitored and reviewed by both the developer and testing teams heads can also called as Defect tracking team.

Chapter 5 Sample ISTQB Questions:

Q1. What is the benefit of independent testing?

    a.   Independent testers are more intelligent than Developers.

    b.   Independent testers can find other and different defects and are unbiased.

    c.   Independent testers identify all the defects.

    d.   Independent testers can test better than developers.

Q 2. The most important task of a Test leader includes?

    a.   Co-ordinate and review the test strategy for the project and test policy for the organization.

    b.   Review and contribute to test plans.

    c.   Analyze, review and assess user requirements and specifications

    d.   Review test developed by others.

Q3. Test Manager should never…..

a. Raise incidents against Documentation.

b. Provide information for risk analysis.

c. Select tools to support testing.

d. Re-allocate resource to meet original plans.

Q4.Which of the following is the biggest cost and time saving by use of CAST?

a. Test design

b. Test planning

c. Test execution

d. Test management

Q5. Which of these is not included in test planning activities?

a. Identifying the objectives of testing

b. Assigning resources for the different activities defined.

c.   Designing Quality plans

d. Defining test approach.

Q6. Match the following Test management terms with their respective activities.

v.   Test planning   1. Test environment availability and readiness

w. Entry criteria   2. Re-prioritizing test when an identified risk occurs

x.   Test estimation   3. Test case execution and defect information

y.   Test monitoring   4. Scheduling test implementation and execution

z.   Test control   5. Calculation of recourses.

a. V-3, w-1, x-2, y-5, z-4

b. V-4, w-1, x-5, y-3, z-2

c. V-4, w-2, x-5, y-1, z-3

d. V-4, W-3, x-2, y-5,z-3

Q 7.If in a particular project, requirements are changing continuously, what should a Test manager do?

a. Should work with the stakeholder of the project and understand the changes in requirement so that he can have alternate plans in advance.

b. Should work on easily workable new requirements into the project and work on difficult requirements later.

c. Both a and b.

d. None of the above.

Q 8. Which of the following helps in monitoring the Test Progress:-

i. Percentage of work done test cases preparation.

ii. Percentage of work done in test environment preparation.

iii. Defect Information e.g. defect density, defects found and fixed

iv. The size of the Testing Team.

a) iv is correct and i,ii,iii are incorrect

b) i,ii,iii are correct and iv is incorrect

c) i,ii are correct and iii,iv are incorrect

d) i,iv are correct and ii , iii are incorrect

Q.9.A project is in execution phase is running eight weeks behind schedule. The delivery date for the software is few months away and as stakeholders are very strict about the timeline therefore the project should be delivered on the delivery date. The quality standards should not be compromised either. Which of the following actions would bring this project back on schedule?

a. Eliminate some of the non-critical requirements that have not yet been implemented.

b. Add more developers from different project team to work on this project to make up for lost work.

c Ask the current developers to work overtime until the lost work is recovered.

d. Recruit more software testing professionals.

Q 10.If you are an experienced project manager and have been transferred to a critical software development project that is in the implementation phase. What would be your highest priority?

a. To make sure that product is delivered on its delivery date.

b. Have a meeting with the stakeholders of the project to establish a relationship.

c. Read the project plan and its main objectives.

d. Change the project plan as per your management style.

Q 11.You are a tester for testing a large system. The system model is very large there are a lot of inter dependencies with in the fields. What steps would you use to test the system?

a. Better and improved vision, more reviews of program.

b. Make a different test plan so that you can test all the inter dependencies.

c. Divide the large system into small modules and test its functionality

d. Test the interdependencies first, after that check the system as a whole

Q 12A Test Manager is managing a very complex project with lot of dependencies on converges at test implementation. If they miss one configuration file then that can lead to meaningless results and that can end up with althea testers of the team to be sitting around for days and repeat the testing. Who is responsible for this incident?

a. Test managers

b. Test lead

c. Project manager

d. Testers

Q 13. You are the test manager and working on system testing. The developer team says that due to frequent changes in requirements they will be able to deliver the system to you for testing 5 working days after the due date. What steps you will take to finish the testing in time.

A. Ask the development team to deliver the system in time so that testing activity will be finish in time.

B. Extend the testing plan, so that you can accommodate the changes.

C. Rank the functionality as per risk and concentrate more on critical functionality testing

D. Add more resources so that the slippage should be avoided.

Q 14.Consider the following for Test approach?

i. Risk based testing

ii. Methodical approaches, such as failure-based .

iii. regression-averse approach.

iv. The size of the testing Team

a) i,ii,iii,iv are true

b) i,ii,iii are true and iv is false.

c) ii,iii,iv are true and i is false.

d) i,iv are true and ii, iii are false.

Q 15. What is true for configuration management?

i.   Developing new test wares.

ii.   Establish and maintain the integrity of the software or system.

iii.   Controlling the version of test ware items.

iv.   Tracking changes of testware items.

    a.   I, ii are correct

    b.   I, ii and iv are correct

    c.   Ii, iii, iv are correct

    d.   Iii and iv are correct.

Q 16. A tester is working on a project to develop an EFPOS system for supermarket. Which of the following is a product risk for such a project?

    a.   More reliable competing product in the market.

    b.   A very high number of defects fix fail during re-testing

    c.   An incomplete system is released in the first version of the test.

    d.   It fails to accept the credit cards.

Q 17. Which of the following statement determines the level of risk?

    a.   The price at which software must be sold.

    b.   Harm to the user.

    c.   Detection of bugs in re-testing.

d. Difficulty of fixing related problems in code.

Q18. According to ISTQB Glossary, Product risk is directly related to……..?

a. The test object.

b. Adverse future events.

c. Potential failure areas

d. A test item.

Q 19.Which one is a Project risk?

a. Failure –prone software delivered.

b. Poor data integrity and quality.

c. Poor software characteristics.

d. Low quality of design and code.

Q 20. Which is NOT an objective of incident reports?

a. Provide ideas for test process improvement.

b. Track the quality of the system under test.

c. To provide feedback about the problem to enable identification, isolation and correction.

d. Determine to test techniques to be employed.

Notes:

6. Tools for Testing:

This chapter deals with various types of Testing Tools that can be useful for Testing in various phases from Test planning to closure. There are so many tools available now a days in market to reduce the effort of testers. Every tool has a specific usage like security testing tools also called pen testing tools, performance testing tools, test management tools, Bug/Incident reporting tools etc. Most of the tools available are licensed ones and some of them are freeware too.

But before using a specific tool in any stage of testing, Testing team must be aware of stuff like Why Automation?, When Automation really required ?, Is the tool really worth to buy or is there any other alternatives ?. These are the questions that testing team must be aware before utilizing the tools for Testing and to achieve better testing within the short time period. This chapter will answer all the above questions and we will discuss about various Tools in Testing and how they are useful and in which phases of testing they are really useful.

6.1.1 Tools classification:

Every tool used in the software testing will have distinct features and characteristics. Some of the tools will also possess more than one feature like a Test management tool will also perform Incident reporting sometimes. Tools can be classified based on their purposes, Freeware or paid tools, Dynamic testing tools or static testing tools, Functional testing tools, Non functional testing tools, Web application testing tools and also can be classified based on the level of testing they can perform like Unit testing tools, integration tools, System testing tools, acceptance testing tools.

There are some tasks that these tools cannot perform and there also some tasks like regression testing, performance testing where we definitely require some tools. This statement tells us that tools are really useful in some cases to reduce the work effort and in some cases human mind can perform effectively than tools. For example to perform Usability, GUI testing tools are not useful in such cases Manual Testing approach can achieve something greater than tools because usability testing is checking the Application under testing can be used by a normal person or not, how the look and feel of application looks like these concepts requires a tester who can think and test the application with user perception, tools cannot perform these of Testing. In other cases like Performance testing usage of tools outnumber the Manual testing process and achieve better results in a quick time where as it is very hard to conduct performance testing manually to check the load testing manually creating a load of 500 users takes time and also wastage of resources and money. So, tool usage is ideal in such stages and manual testing is better in some other stages.

But there are some certain limitations of tools and tools testing of certain function in software may also show some side effects on the other parts/modules of the software which is called Probe Effect. For example a tool performing Database testing have to interact with the database to execute every test case and condition, with this interaction it may takes some time and this interaction time also varies with the tool we are using some other tools can perform better than the current tools. There is also uncertainty in tools all the tools cannot process the same result in same time some tools will identify the critical conditions while some other tools are not capable to do so. For example Selenium Web Driver tool can perform Web application testing better than any other tools but incapable of Desktop and Mainframe application testing like QTP tool. So every tool will have different features, similarities too. Some tools will use their suite of tools while some other tools have only certain limitations in usage for specific purpose Bug Zilla tool can only be used for Incident reporting it cannot perform Test management and Test designing activities.

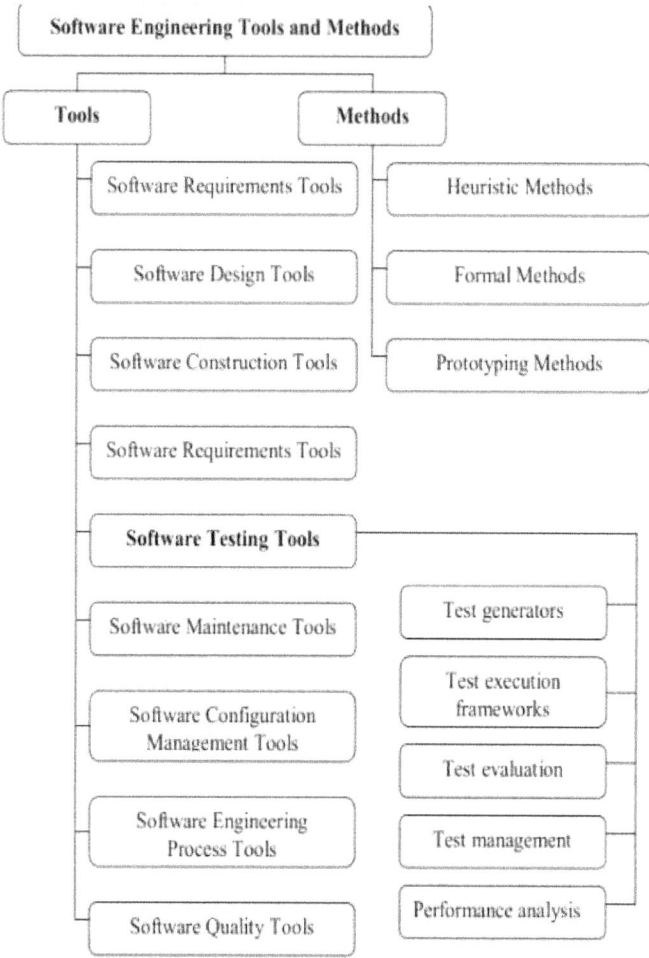

Figure: Classification of Testing tools (Swebook, 2004).

The above describes the classification of testing tools in every phase of testing like in requirement Analysis phase we use requirements tools which are capable of covering all the requirements in the project and also alerts if any requirement is missed in testing. There are also some other kinds of tools like configuration management tools, test designing tools capable of generating test cases and scenarios. We will look briefly about the tools and their specific usage in further sections.

6.1.2 Tools for Management of Testing:

Tools that can manage the tests or tools that can manage the testing process are called as Test management which can be at any phase of testing. Hence these are the first tools that can be used in early phases of testing. These management tools can perform both the tasks like managing the testing process and testing. In fact Test management tools can be used either by the Test leads or Test manager at any stages of testing.

Test management tools

There are lots of Test management tools available in the market these diversification introduces complexity in choosing the tools. For example the same vendor will have different types of tools supporting distinct characteristics and features. so choosing a test management is a complex process. Let us find the common features that any Test management tool must support so that we can use a checklist to decide which is of those tools are right fit for our case. The below list provides the common features that any test management tool should possess:

- Managing Tests (Test plan creation, Test strategy creation and updating, managing the Test cases and   keep track of scenarios, No of tests performed, passed/failed tests, coverage etc).

- Scheduling the future tests to be executed.

- Management of Test effort and test activities like time spent for Test execution and test designing etc.

- Better interaction with other tools like incident management tools, designing tools, coverage tools, screen capturing tools when required.

- Summarize the results of tests from test execution tools.

- Must be flexible and easier to use

- Preparing interactive reports capability with graphs and pictures representing no of tests passed and failed, Test coverage percentages etc.

- Capability to share and maintain testing activities in one common place which can improve the collaboration between the testers in a team of company.

- Must support Additional plug-in facility to connect with other tools for automation and performance testing.

These are the features that must be supported by any Test Management tool. Most of the test management may not support all the above mentioned features too. Let us know the names of popular test management tools available in present market.

Examples: IBM Rational Quality Manager, Zephyr, HP ALM Quality Centre, Borland Silk Central, Qmetry, Test Rail, Test Lodge, Qtest are some of the Test management tools.

Requirements Management Tools

Requirements management is very essential initial process for the success of any kind of project. Especially this Requirement Management is really useful when we are dealing with very large projects with thousands of requirements. Missing requirements specified by the user in such huge projects is quite common activities to avoid such conditions Requirements Analysis tools are required.

Requirement tools are capable of keep tracking many requirements in a huge project. Some consider Requirement Analysis tools are not Testing tools of course they are not fully fledged testing tools but in very large projects testers and Business analysts can easily ignore some customer requirements which can show a great impact on the quality of the project.

Let us look at the common features that requirement tools should support:

- Storing the requirements in a clear and concise way

- Ability to store the attributes related to requirements

- Easier requirements input mechanism for the user to include requirements in to tool

- Reuse facility of common requirements to use for other similar projects

- Requirements validation capability to check the quality of requirements included by user

- Decision making capability to prioritize the requirements

- Requirement Management tools should possess the requirements coverage ability with percentages and interactive graphs

- Ability to update the changed requirements

- Alert functionality with emails to the concerned person when major requirements are missed and ignored

- Interface facility with several testing tools and common usage tools like word pad and spreadsheets etc

Examples of Requirement management tools:

IBM Rational DOORS, Sparx solutions, iplan, requirement one, Relatics, Blueprint Requirements centre, Caliber RM, Jira Agile etc.

Incident Management Tools

Incident management tools are really useful for the testing team to report the incidents they found during testing and to keep track of incidents reported from the starting of a project in order to find and fix the incidents. Primary job of a tester includes finding/reporting/recording/tracking/managing the incidents. Usually small companies use excel spread sheets as their Bug reporting and bug tracking tool but as the size of the projects, the number of test cycles, and the no of the people involved in project grows – it becomes absolutely important that we need a sophisticated software which can take care of reporting, managing and tracking of incidents efficiently. So, that tester can concentrate more on actual testing rather than reporting and managing the incidents already found.

Incident management tools are also known as Bug reporting, bug tracking, and bug management or defect tracking, defect reporting and defect management tools. However we call it but using a bug tracking system or app within your project will help you leverage from an effective and better quality end product. However, choosing a right bug tracking system is not at all an easy-to-go task. It requires you to consider a lot of things based on your project requirement.

Let us look in to the common features that any incident management tool should support.

- **Reporting facility** with all the fields required about the incident, environment, module, severity, screenshots etc.
- Status and Assigning capability to assign the incident to the concerned person and status of the incident.
- History/work log/commenting facility for everyone in the team about the bug Ex: chat box to communicate within the team about the bug
- Attachment facility to attach graphs or charts, videos, URL's if needed
- Description of Every bug must be clearly classified under which module, which web page the incident is occurred will help developers to reproduce the incident.
- Alert facility through email or SMS when most critical show stopper incidents found.
- Multiple project support

A part from all the above mentioned there are many that can be included in a incident management tool like capability to vote the bugs and capability to export the bug reports in to various formats like PDF, .doc etc. But the above mentioned features are really necessary for any kind of incident management tool.

Examples of Incident management tools:

BugZilla, Atlassian JIRA, IBM Rational Clear Quest, Bugherd, Fog Bugz, Redmine, eTraxis, You track etc

Configuration Management Tools

Configuration management tools are not the testing tools but proper configuration will ensure better testing and development too in the project. A Tester uses various types of software's and browsers and hardware's to test the application from the beginning of the project. Apart from that there will be several builds of code from the initial days of project it is very important to test the software in the same hardware and software conditions to check the quality.

In some cases end user may require his application to work in various versions of browsers and operating system. To manage and configure all these software builds, version we need a configuration management tool in the project this can   also be done without using any tools but if the size of project grows it also increases the complexity to manage.

For example we have developing a web application and testing is also done in various kinds of Operating systems like windows, unix etc. But after the delivery the software is not working well with MAC os and we have missed the customer requirement of Application should work for any OS and any software. To avoid such conditions we require configuration management tools.

The basic features that a configuration management tool should support are:

- Keep track all the versions of software, hardware and browser versions used for developing and testing from the initiation of project.

- Software builds and patches management and tracking.

- Take requirements from specification to testing

- Configuration status account reporting

- Integration of software's used for testing and developing

Examples of software Configuration management tools:

Mercurial SCM, Bazaar, IBM Rational Synergy, smart frog etc

6.1.3 Tools for static Testing:

As we discussed in the earlier there are two main processes in testing called verification and validation or static testing and dynamic testing.

In this section we will discuss about the static testing tools that can be helpful for the walkthroughs, reviews in static testing.

Review tools

Review process is familiar in static testing where many people in different locations will gather for a formal meeting and share their views regarding the project in different aspects by reading the documents related to the project. Especially in such cases review supporting tools will help for a better reviews because everyone don't have same skills in reading and understanding the project documents in detail. It is very important to make all the suggestions and changes specified by the people in review meetings. We can also use spreadsheets and word documents for doing this activity but the customized tools developed especially for reviewing will plan and organize in a better way than spreadsheets and documents.

Review process support tools can be used in both formal and informal reviews as well. let us look at the common features that review tools should support.

- Unique id, date for every suggestion and comment made in review process to avoid the future resemblance conflicts

- Better communication between the people involved in reviews located at various locations

- Keep track of comments and issues found from the first review

- Monitoring the reviewing status whether the review is passed or failed

- Clearly explain the Objective and aim of the review with description of related documents used in the review process

Static Analysis Tools

Static analysis is also known as source code analysis is a white box testing technique and static analysis tools run the static code with an aim to find the possible vulnerabilities in the code are mostly used by the developers. Static analysis tools checks the software in non run time i.e. without executing or running the code. This testing will ensure to find the earlier flaws inside the software and will also help for the code coverage to the developers. Static code analysis also strengthens the security of our application by detecting the flaws in the code and will also suggest the changes that can reduce the complexity of the code so that which saves the execution time and server load as well. Static analysis helps to deliver the better quality code to the QA department so that test can be done more effectively.

A QA tester will test the functionality and performance of an application under expected operating conditions and unexpected conditions as well, while static analysis tools will look at the code from a technical perspective and identify issues that might be missed based on the parameters used to structure the QA test program.

The following are the common features that a static code analysis tool should possess:

- Ability to find syntax, lexical, semantic errors in the code.

- Ability to find the code coverage

- Ability to provide recommendations on the code to developer for better use of resources and to reduce the code complexity.

- Should support static code analysis for multiple high level languages like C, C++,C#, Java, Ruby, and Perl etc.

- Ability to calculate the software metrics from the code.

Examples of static Analysis tools:

HP Fortify static code Analyzer, veracode, visual studio team system, Microsoft Pre Fast etc.

Modelling Tools:

Modelling tools helps to stream line the process of development and designing in a project. Modelling tools helps to choose the model of software being developed. Especially these modeling tools are used by the designers and developers in designing phase to choose the designing architecture of software. Modelling tools validates the object models and state models.

Static Analysis and Modelling tools are used in the initial phases of developing and designing a project to find out the flaws in earlier stages. Because it is always easier and costs less to fix the bugs founded in earlier stages of a project. Modelling tools identify the test conditions and scenarios using any kind of modeling language like UML. Modelling tools helps to find the flow of data and flow sequences architecture of a project.

These are the common features that any Modelling tolls should have

- Ability to integrate model artifacts from business requirements phase to implementation phase.

- Identify and prioritize the areas of modeling

- Ability to import and export from resources like word, excel, xml, csv and pdf files

- Validate the defects and flaws within the model

- Ability to Analyze, design, view and edit the models in various abstraction levels.

- Decomposing the project in to several easier modules to design the artifacts.

Several modeling tools comes as a plug-in to integrated development software's like Eclipse and Net beans and some of them are starUML, AndroMDA, BOUML, Papyrus etc.

6.1.4 Tools for Test specification:

Tolls discussed in this section can be used for various Testing activities like Test designing which involves Test cases and Test scenarios generation and execution, Test data preparation and for various testing techniques etc.

Test designing tools:

Test designing tools reduces the major work of testers in creating test cases and scenarios. Test Designing tools have the capability to design test cases and test scenarios with the requirements.

Test designing tools have the capability to measure all the possible combinations of testing in a project. For example in a project to test new registration functionality in registration module test designing tool will discover all the possible combinations and the ways to test the fields in it. Some tools will also help to identify the test coverage of the project which helps for the faster test case generation and testing process, more coverage's per test and more thorough test plans to find more bugs. Test designing tools also have the capability to compare the expected results and actual results while executing the Test cases. Anyway these tools can't replace the Tester job because tools will only have the capability to perform according to the inputs given by the Tester. Tester should control the tools but never tools replace the testing job they will help to achieve the testing more effectively, accurately and quickly by saving time and effort.

The Common Features that any Test designing tools should have:

- Ability to generate the Test inputs from the requirements, Test conditions, and Source code and modelling tools.

- Ability to generate formal Test cases and Test scenarios.

- Ability to calculate Testing metrics and Test coverage of the project.

- Ability to generate and compare the expected outputs and actual outputs from the inputs

Examples of Test designing tools:

Hexawise, All pairs, Pair wise Testing, PICT web shell etc

Test Data Preparation:

To write test cases tester needs input data from predefined locations like databases. Generating test data is a key task in testing especially in testing projects which involves testing with huge volume data requires Test data preparation tools usage. By using test data preparation tools testers will have all the test data and possibilities to test so that they can ensure that no part of the application is missed.

Test data preparation tools allows you to fetch data from the existing database to create, generate, manipulate and manage the data to be useful for the process of testing. To perform Data volume testing and performance testing these types of tools are useful. Any important test case will not be missed by improper testing and incomplete testing environment with the usage of Test data preparation tools. Ability to fetch data from any kind of sources and formats like doc, xml etc and to make the data useful for software testing is the main feature that a Test data preparation tools should support.

6.1.5 Tools for Test Execution:

Test execution tools execute the tests in the software testing process, test execution tools are also known as test running tools. Usually test execution tools will need some coding knowledge to write test scripts. Test scripts are the test cases and test scenarios in terms of automation testing. Automation test engineers who are capable of generating test scripts will use those tools in the process of regression testing. There so many testing tools available in the market with distinct features like functional regression tools and web application automation tools etc.

Test execution will offer two kinds of facilities for test execution; the primary feature is called record/playback or capture/ playback tools which are capable of recording the tests just like a record button in mp3 players. Automation testers can view their tests whenever requires by using the playback functionality. Recording feature enables to record step by step by testing process and will save those results in various formats for future references. But in the projects which are very dynamic with changing requirements record playback tools are not suggestible because recorded scripts are no more useful for the modified modules and it takes time to modify the recorded to suit the updated requirements.

The other feature of test execution tools are descriptive programming which executes the tests with scripting languages. Automation tools will create test scripts using scripting and programming languages so that test execution tools can interact with the Application under testing (AUT) to execute the tests. Several test execution tools also support key driven and data driven approaches to promote the repetitive testing tasks in regression testing to test the modifications and the impact of modified over the remaining modules in a project. Descriptive programming feature enables test execution to perform repetitive tests with various inputs and combinations quickly. For example with the descriptive programming test execution tools we can test the registration form of any application several times with various inputs by using a single program with loops written in supported programming language.

But these tools also have some limitations like some of the tools are only capable for Web applications testing they can't execute desktop and Mobile applications. Most of the testing tools have limitations in supporting the programming languages written by automation testers for test execution. For example QTP tool of HP company only supports VB Script language to write test scripts but capable to test various kinds of applications like Web applications, Mobile, Desktop and Mainframe applications. Let us look at the common features that test execution tools will support:

- Ability to record and playback the manual tests of any kind of application.

- Ability to export and save the recorded test results in various formats.

- Ability to differentiate and log the test results pass/fail conditions

- Interaction with other tools like Test management and configuration management tools

- Ability to test any kind of applications like web, mobile and desktop applications.

- Should support various languages for writing test scripts.

- Ability to interact and synchronize with application under testing

- Ability to fetch test data and inputs from word documents and spreadsheets excel documents for data driven testing.

- Automatically waiting should be enabled before fetching each input in testing

Examples of Test Execution tools:

HP QTP, UTF - (Allows recording and descriptive programming facility to test any kind of applications like Web applications, desktop and mobile applications).

Selenium IDE- Recording and playback tool for only Web applications in Firefox browser.

Selenium Web driver- Descriptive programming tool which allows Automation test engineers to write scripts in various programming languages but supports only Web application testing and not able to test desktop applications.

SOAP UI- only supports Web services testing.

Unit testing Tools:

Unit testing tools are used by the programmers in the phases of unit testing and integration tests to test the components individually. In Integration testing programmers will use these tools to interconnect and integrate several independent stubs and drivers to form a module of program. Tools will support various integration approaches like top down and Bottom up approaches to integrate several independent modules as discussed in earlier sections.

These tools especially useful in Agile models to carry out the unit testing, integration testing parallel to the development in a project. Unit testing frameworks have been developed in unit testing tools to simplify the process of unit testing. Unit testing tools also provide code coverage functionality also but unit testing tools are specific for every programming language the system developed.

Examples:

Junit- only supports java written programs unit testing, Nunit- only .net applications unit testing, JSunit- java script unit testing, CPP Unit- c++ unit testing.

Test comparators:

Test comparators help to see what the exact error is caused for failure. Test comparators compare the expected output and actual output of test results. Testing process involves the comparison of actual results to expected results to find out the errors in a project this is what exactly the test comparators will do. Test comparators uses two techniques to compare the test results they are dynamic comparison and post comparison. In dynamic comparison Test comparators will compare the Actual test results with expected test results by executing the tests. In post comparison comparators will compare the results after the test execution.

 Security testing tools:

Security testing tools are also known as penetration testing tools or pen testing tools used by the testers to assess the security mechanism of Application under testing. These tools will help to find the security

---

flaws in the application especially in banking and financial applications security is very important before facing any security breaches. Security tools will helps to find the data security functionality, encryption standards of sensitive data stored in the application. Security testing tools will also help to test our application against several security attacks like Denial of service, SQL injection, Cross site scripting attacks, phishing attacks and also checks for authorization and access control issues in application like Banking user will only have limited access and Admin have some additional access powers. With the help of security tools one can find the vulnerabilities in the application that can cause security attacks and failures. Several security testing tools are there a part from them several browsers supports add-on's for specific security check like SQL injection and brute force attacks etc.

Example tools:

Wapiti, skip fish and chrome add-on's like form fuzzer, sql injector, tamper data etc

Performance testing tools:

Performance testing tools helps to find the performance of application under testing in various loads. These tools mainly check the applications by creating various virtual user loads to test the application. Performance testing tools mainly saves money and time compared manual performance testing which requires great resources to create loads. Performance tools can test Load testing and also able to create stress to the application under testing by extra load to check the AUT behavior under stress this process is also called as stress testing. Some of the performance tools help to check the endurance testing to check AUT behavior under repetitive load for overnight and days together. There are specific licensed performance Monitoring tools and some of the other cloud based web applications will also measure the performance with an input of URL from the user.

Performance testing tools also requires some programming knowledge to automate the performance testing process. These tools also have the capabilities to measure the time taken for responses (response time for communicating with server in various loads of traffic) in the Application. Performance is critical for some Time based applications where these tools can help a lot. The common features that performance should require are:

- Ability to measure time responses precisely

- ability to create virtual loads

- Ability to assess the peak load or server crashing load limit

- supporting various interactive graphs to show the loads and responses time

Examples:

HP Load runner, Jmetre, Load Ui, Load complete and web applications like Gm Matrix, Nuestar, Alertsite etc

Monitoring tools

Monitoring tools continuously monitors the systems and alert the users before any unwanted things like failures arise. These tools may monitor the network, websites, servers, databases, internet usage of employees in a company.

Usually these servers and databases are very costlier so by using any specific kind of monitoring tools we can avoid the future failures. For example network monitoring tools like stat seeker will help to monitor the network continuously and alerts the user through email and SMS when any component in the network like switches, routers fails to operate. These primitive alerts can prevent from future failures which are usually costlier to deal with. The common features of Monitoring tools are:

- Identifying the future failures and alerts the user through emails, SMS

- Ability to save the historical past log data

- Utilizing the resources in network or servers optimally

- Monitoring the real time user traffic for the legimate users and duplicate users

6.1.6 Other Tools:

Apart from the above tools discussed in this chapter testers will also use several others tools and add-on's in their daily activity to make their testing easier. Basic tools like spreadsheets and word documents which are used for Test designing can also be considered as tools. Apart from them some testers will also use online Mind maps tools to create test plans and Bug advocacy. Screen recorders and screen capture tools are also used in incident reporting to capture and highlight the errors in web screens of an Application under testing.

HTML validators can also be used to check the HTML validations of web applications these validators widely available in online resources. Some Monitoring tools are also used to check the vulnerabilities and SEO optimization of web applications. Mobile emulators are also used to test mobile applications where we can run our application in any type of mobiles widely available in internet. Broken link checkers add-on's used to check the broken links in any web application. Sticky notes software's which allows testers to make some quick notes and to add some to do lists, checklists etc.

Any tool that can reduce the tester work can be considered as a non testing tool or other tools.

Examples of other tools that can help testers in their testing journey:

- Qsnap, snagit- screen capturing tools.

- Evernote, simple sticky notes- Stick notes software's

- Robotium, Test object- Mobile testing tools

- http://www.brokenlinkcheck.com/, http://validator.w3.org/checklink- Broken link checkers

6.2 Effective usage of tools: Benefits and Risks

In this section, we will look at the effective usage of tools in testing. Every tool has risks and benefits too. Licensing costs for some tools are very high so we have to assess before spending money on tools whether they are really useful for our testing and is there any alternatives to them.

It is always better to know when to go for automation (Testing using tools) and when manual testing can give better results. It is not suggestible to go with Automation testing from the beginning of project. Most of the test execution tools will only help in regression phases of testing so before taking a decision towards tool and spending money we also have assess the risks and benefits and the tool impacts on testing.

Let us consider this scenario, A company decided to deliver the project as quickly as possible and opted for a licensed Automation tool which costs a 1year of a test engineer salary. But the existing Testing department are not familiarized with Automation testing so again they have to recruit an Automation tester and they have train their all existing employees for the effective usage of new tool. In this scenario after considering the training costs to employees and recruiting costs, licensing costs will become higher than the allocated budget for testing. Moreover newly trained employees will not have any real time experience in using tools which affects the quality of project.

In some cases like regression testing as we discussed before Automation tools will helps a lot. Let us look the benefits of Automation and tools:

Benefits of tools:

- Ability to run test cases repetitively-suitable for regression testing

- Increases accuracy and efficiency of testing by eliminating human errors

- Time saving compared to manual testing

- Increases reusability of tests developed through automation tools for future projects

- Automation tests can be run simultaneously over several machines in same time

Risks of tools:

Along with the benefits there are also some risks and disadvantages of using automation tools they are as follows:

- Limitations of tools- some tools will perform only limited operations like Load runner will perform only Load Testing they are not capable to perform functional testing.

- License costs- in some cases as explained above automation testing will cost more than manual testing

- Underestimating the effort and time required for maintaining and writing test scripts- Requires high skilled professional team for Automation.

- Depending totally on the tools- Some free tools like selenium will not offer or guarantee any support services if something goes wrong so totally depending on the tool is not safe game.

- Test maintenance costs are very high with the usage of record playback tools particularly in the case of changing requirements.

- Maintaining the test data is also difficult, complex Analysis is also important if any Automation test failed.

Tool Selection criteria:

There are varieties of tools which perform variety of activities during test process. Even though there are more than 600 test tools available in the market there is no specific criteria to select a tool. Before selecting a tool we should clearly know the capabilities and limitations of the tool considering the costs involved for training, maintenance and licensing as well.

Primarily if the tool is really required and we can't test those specific conditions manually or testing manually requires more effort and time then we have to choose for automation tolls by comparing the available popular specific tools that can justify our company testing needs.

The Most important in choosing a tool:

- Functionality: Accuracy, Interoperability, security and compliance

- Reliability: Fault tolerance, Recoverability

- Usability: understandability and operability for basic users

- Maintainability: stability and adaptability

- Vendor support: quality of support, level of support, warranty, regularity of updates

- Licensing costs: Group license costs, Enterprise license costs, Freeware or commercial, price consistency and license expiry conditions and usage limitations because some tools offer 1 year license, lifetime license and some other will give access to limited hours.

The above mentioned criteria can be used as a check list in choosing a right tool for saving money and reducing testing effort at the same time to increase the accuracy of testing.

6.3 Introducing a tool in to an organization

Just buying a testing tool and introducing the tool in an organization to start testing will not guarantee any cost saving to the company. It is a process to select a tool and to introduce it in to the organization. Every organization will have their own testing process and before introducing any tool we have to clearly understand the existing process of testing in any organization. Even though test automation can bring benefits, we need to analyze how the new tool and Automation testing will affect the company's current testing process and testing staff.

"Any tool is intended to make the testing process effective and efficient but there also risks most of risks in tools are concerned with over expectations and over dependencies. Before introduction of any tool in organization we need to define our goals that we wants to achieve with the introduction of tools. How test teams introduce an automated software test tool on a new project is nearly as important as the selection of the most appropriate test tool for the project. A tool is only as good as the process being used to implement the tool. It's worth the effort to invest adequate time in the analysis and definition of a suitable test tool introduction process. This process is essential to the long-term success of an automated test program. Test teams need to view the introduction of an automated test tool into a new project as a process, not an event. The test tool needs to complement the process and not the reverse. In other words, the tool is only as good as the process, meaning that a process has to be in place before a tool can be brought in".(Dustin, 2001).

The primary operations that any organization's testing team has to do before introducing any tool are:

- Evaluating the capabilities of tool

- Evaluating the existing testing process

- Evaluating the future risks with the introducing of tools

- Evaluate the cost of capital investment and operational costs with the tool

- Check the tool compatibility with existing software's and hardware's

- Reviewing the requirements of tool and project schedules

- Assign the roles and responsibilities for testers when new tool is introduced

- Determine the technical skill set of existing team to utilize the new tool

After completing the test tool consideration phase based on the above specified operations, then testing team can perform the final analysis necessary to support a decision of whether to commit to the use of an automated test tool for a given project or not by conducting a pilot project with the tool.

Pilot Project:

Pilot project is the informal of checking the tool capabilities and impacts in an organization. After the tool is introduced in to organization it should not be used immediately in the real projects because if something goes wrong it will impact the whole project. So before introducing the tools in to the real project we have carry out some operations on the tool with sample projects like creating datasets and using all the features in a tool and assessing the tool limitations, creating and testing data sets of sample projects, checking the effectiveness and accuracy of the results tested by the tool. We can even note the time period taken for sample project testing with the tool so that we can evaluate the real benefits of tool when it is introduced in to the real project. This pilot project will also help the existing testers to be familiarized to the new tool so that in mere future they can operate the tool more accurately and efficiently.

Chapter 6 Sample ISTQB Questions

Q1which tools help to support Test Execution and logging?

    a. Coverage measurement tools, security testing tools, Unit test framework tools.

    b. Coverage measurement tools, monitoring tools, data quality assessment tools.

    c. Security testing tools, Test comparators, modeling tools.

    d. Security testing tools, Unit test framework tools, monitoring tools.

Q 2 Which test activities are supported by Test harness?

    a. Test management and control.

    b. Test execution and logging.

    c. Test specification and design.

    d. Performance and monitoring.

Q 3. Which statement is False for Test harness?

a. Exercise software which does have a user interface

b. It is used to run automated tests or comparisons.

c. Often custom-build.

d. These are used as simulators.

Q4. Which should be an objective to introduce a testing tool in a pilot project?

a. To assess the competence of a tester in using the tool.

b. To assess whether we can achieve benefits at low cost.

c. To complete the testing of the project

d. To discover the requirements of the tool.

Q 5. Find the mismatch?

a. Modelling tools- validate software models by enumerating inconsistencies.

b. Configuration management tools- used for storage and version management of testware.

c. Incident management tools-Provide interfaces for executing tests.

d. Requirement management tools – Enables individual tests to be traceable.

Q 6. A Performance tool would be able to perform all of the following except?

a. To check memory leaks

b. Monitor and report on how a system behaves under lot of load pressure.

c. Verify and report on usage of specific system resources.

d. Evaluates the ability of software to protect data confidentiality.

Q 7.Which of the following tools are used by only developers?

i.   Unit test framework, coverage measurement, review tool

ii.   Static analysis, dynamic analysis, coverage measurement.

iii.   Dynamic analysis, data quality assessment, review tools

iv.   **Unit test framework, static analysis, dynamic analysis.**

a.   i.,ii are correct iii , iv are incorrect

b.   ii, iv are correct and I, iii are incorrect

c.   i., iii, ii are correct iv is incorrect.

d.   iv is correct, I, ii, iii are incorrect.

Q 8.A test manger decides to use functional test execution automation tool in his project. The requirements are expected to change frequently. Manager puts some Of the manual testers through training program on how to use the tool. Which of the following is likely to be true?

a.   Automation is likely to fail because of frequent changes and lack of experience.

b.   Automation is likely to fail because it is not the right way to automate in this situation.

c.   Automation is likely to succeed because automation is very useful for frequent changes

d.   Automation is likely to succeed because the team has been trained on tool.

Q 9.Select the test specificaton tools?

a.   Monitoring tool, load testing

b.   Security , Test comparators

c.   Test design, Test data preparation tools

d.   Review, modelling

Q 10. Which of the tool is used for testing non-functional quality characteristic?

a.   Test execution

b.   Test Management

c.   Performance

d.   Static analysis

Q 11. What are the benefits of using tools in software testing?

a.   Reduction in number of testers and better objective

b.   Better quality of software and reduction in repetitive work.

c.   Reduction in documentation and better quality of code.

    d.   Objectives are not necessary to define.

Q 12. What can be risk in using tools in software testing?

    a.   Unrealistic expectations from the tool.

    b.   Tools may find defects that are not possible.

    c.   Tools will repeat the same steps again and again.

    d.   No reliance on tools.

Q 13. Which of the tools is used in automation of regression testing?

    a.   Capture/Playback

    b.   Test comparators

    c.   Modelling tools

    d.   Data quality assessment.

Q 14.What is NOT one of the characteristic of security testing tools?

    a.   Identifying virus.

    b.   Simulating various types of external attacks.

    c.   Checking memory leaks

    d.   Probing for open ports.

Q 15. What should be an objective for a pilot project for a new tool?

    a.   No objectives for using a tool

    b.   To decide on standard ways of using the tool that will work for all potential users.

    c.   To check if it can overtake all the manual effort.

    d.   To ensure that it is beneficial at any cost.

Q 16.which tool carries out analysis of the source code without executing the tool?

    a.   Dynamic analysis

    b.   Static analysis

    c.   Test execution

d.   Review tools

Q 17. Which of this can cause failure in implementation of Test Tool?

a.   High Price of the tool

b.   Underestimating the use of tool.

c.   No defined requirements for the tool.

d.   High maintenance of the tool.

Q 18. Which defect can be found by "Test harness"?

a.   Memory leaks

b.   Security characteristics.

c.   A defect in the middleware.

d.   Data conversion.

Q 19. Which of the following statement is FALSE?

a.   In a keyword driven testing approach, the spreadsheet contains keywords describing the actions to be taken and test data.

b.   A data-driven testing approach separates out the test input, usually into a spreadsheet.

c.   Static tools applied to source code can never enforce coding standards.

d.   Test management tools need to interface with other tools or spreadsheets in order to produce useful information.

Q 20. What is one of the types of scripting technique for test execution tools?

a.   Key-hole driven

b.   Key-word driven

c.   Data sheet-driven

d.   Data- quality driven

Answers to ISTQB Questions:

Question No.	Chapter 1	Chapter 2	Chapter 3	Chapter 4	Chapter 5	Chapter 6
1	a	b	a	a	b	a
2	b	c	c	b	a	b
3	a	d	c	a	d	a
4	c	c	b	c	c	b
5	b	d	c	a	c	c
6	b	a	d	b	b	d
7	d	d	a	d	c	b
8	c	c	c	a	b	c
9	a	d	b	c	a	c
10	c	d	c	c	c	c
11	c	b	b	c	a	b
12	a	a	a	a	a	a
13	c	c	c	c	c	a
14	d	d	c	c	b	c
15	d	c	b	a	c	b
16	a	a	d	c	d	b
17	a	a	d	c	b	c
18	c	b	b	c	a	c
19	c	d	c	c	d	c
20	a	a	c	c	d	b

References:

Black, R., (2011), What is Test Control? , Retrieved from:

http://www.rbcs-us.com/blog/2011/04/04/what-is-test-control/

Dustin, E., (2011), Introducing Automation testing to a project. Retrieved from:

http://www.informit.com/articles/article.aspx?p=21479

Graham, D., veenendaal, E.V., Evans, I., Black, R. (2011). Foundations of software testing- ISTQB Foundation Certification.

Montvelisky, J.,(2009), The simple differences between project risks and product risks, Retrieved from: http://qablog.practitest.com/2009/01/the-simple-differences-between-product-risks-project-risks/

Swebok, A., (2004), IEEE Guide to Software Engineering body of Knowledge.

Useful Resources for future Test Engineers:

Testing Purpose	Online Tools	Chrome Add-on's	Mozilla Add-on's
Cross Browser testing	www.browserstack.com www.browserling.com www.browsershots.org	• BrowserStack Local • Test IE	Crossbrowser
Html Validators	http://validator.w3.org/ http://html5.validator.nu/ http://www.onlinewebcheck.com/	• HTML Validator • Validity • Kingsquare HTML Valdiator	• W3C Page Validator 3.0.0
Broken Link checkers	www.pingdom.com http://www.brokenlinkcheck.com/	• Simple Test Io Site checker • Check My Links	• LinkChecker • Pinger • Total Validator
Screenshot Capture	http://awesomescreenshot.com/	• Q snap • LightShot	• Fireshot • Aviary screen capture
Performance Testing	http://www.loadui.org/ http://gtmetrix.com/	• Page Speed Insights • BlazeMeter • Speed Tracer	• Extended status bar • Y Slow
Screen Resolutions Testing (GUI)	http://quirktools.com/screenfly/ http://www.screen-resolution.com/	• Window Resizer • Screen Resolution Tester	• Fire resizer • Window Resizer • More Display Resolutions
Cookies Testing	http://www.html-kit.com/tools/cookietester/	• Testing Cookie API • EditThisCookie	• Edit Cookies • View Cookies
Mobile Application Testing	www.testobject.com www.silkwebmeter.borland.com www.perfectomobile.com	• Ripple Emulator Beta • Mobile/RWD Tester	• Firefox OS 1.1 Simulator • Full Screen mobile
Penetration Testing tools	https://www.websecurify.com/ https://www.scanmyserver.com/ https://www.netsparker.com/web-vulnerability-scanner/ http://ophcrack.sourceforge.net/ http://www.openwall.com/john/	• D3coder • Form Fuzzer- automatically fills forms • XSS chef- checks xss vulnerabilities • Behind the Asterisk- reveals masked passwords	• SQL inject me- Checks for SQL injection attacks • Tamper Data- modifies http/ https • HackBar • URL Flipper

Good Testing Resources for future QA Engineers:

It takes life's to learn from our own experiences. It is always a smart people's choice to follow the Leaders to learn tips in testing, tactics and their experiences will also help us to become Testing Leaders in future.

Blogs to be followed:

1. Pradeep Soundararajan – TesterTested
2. Parimala Shankaraiah – CuriousTester
3. Sharath Byregowda – TestToTester
4. Ajay Balamurugadas – EnjoyTesting
5. Manoj Nair – TestingRedefined
6. Mohit Verma – TestingNook
7. Santosh Shukla – ProudTester
8. Vipul Kocher – Eternal Student
9. Dhanasekar S – Pragmatist
10. Nandagopal R – Testing My Way
11. Madhukar Jain – A Rapid Tester
12. Shrini Kulkarni – Thinking Tester
13. Mandeep Singh – Testers Planet "A creative destruction"
14. SaiPavan Gubba's blog
15. Krishnaveni - Passionate Tester
16. Amit Kulkarni – Bug Teaser
17. Deepthi J C's Blog
18. Bharath S – Ticket Number
19. Jigar Patel – Quick Start for Beginner
20. LN's Testing Blog – TESTER by INSTINCT, not by CHANCE.
21. Manjinder's Blog
22. Sudhamshu's blog – A Sane Tester
23. Shiva Mathivanan's blog – A Sapient Tester
24. Mansi Rao's blog – Testing Summit

25. Yagnesh Shah's blog – A budding tester

26. Sunil Kumar's blog – Vivid Tester

27. Malini Mohan Kumar's blog - MaliTheNerd

28. Prasanna Kumar Sabat's Blog - PrasannaSabat

29. Deepak Malladad's Blog – Deepak Malladad

30. http://selenium-by-arun.blogspot.in/ -selenium

31. Guru99.com

32. Softwaretestingclass.com

33. Softwaretestinghelp.com

34. Teatimewithtesters.com

35. Softwareqatest.com

36. Testingthoughts.com

37. Exploringuncertainity.com

38. Testalways.com

39. Softwareqatest.com

40. Testingexcellence.com

41. Softwaretestingportal.com

42. Softwaretestinggenius.com

43. Softwaretesting.net

44. Opensourcetesting.org

45. Blogqatestlab.com

46. http://learning-selenium-webdriver.blogspot.in/

47. Targetprocess.com   (Testing at the edge of chaos)

Crowd Testing Sites:

1. www.99tests.com

2. www.utest.com

4. www.passbrains.com

Test Gurus to follow in Twitter:

- James Bach
- Martin fowler
- Cem kaner's Blog
- Misko hevery
- Sara ford"s weblog
- Santosh Tuppad   (tuppad.com)
- Tester's circle
- Testing ninjas
- Rhythm of testing
- Swtester.blogspot.ch   (Lessons learned by a Tester)

Tips to Pass ISTQB certification Exam:

ISTQB Certification is standard qualification for the professionals in QA industry from the beginner to   Testing expert level ISTQB offers various certification paths like

- Certified tester foundation level CTFL
- Advanced Test Analyst
- Advance Technical Test Analyst
- Advance Test manager

Certifications will definitely make you to stand apart from other candidates in your interviews and also makes you to reach higher steps in your career path. Here are some tips to get passed in your ISTQB CTFL certification exam

- Study the syllabus and Glossary terms of ISTQB clearly
- Have a clear understanding of basics in software testing from the requirement phase to Test closure.
- Have a good command on Test implementation and Testing Models.
- Make use several online resources along with this book
- Give some mock tests regularly before the real exam

- Read and solve special section like branch, path, and statement coverage questions. Try to understand the concept and you will be easily able to attempt the questions correctly.

- Gather some actual questions, practice sample papers and try to solve them within the exam stipulated time. 40 questions in 90 minutes for each ISTQB Foundation mock exam.

- Try to find out how many questions you attempted correctly in the mock tests. ISTQB Foundation pass mark – you need to score more than 26 questions correctly out of 40 questions.

Study with Perfect Plan, confidence and positive attitude WE WISH YOU ALL THE BEST.

WHO WE ARE:

GetSkills is social initiative started by a group of working IT Professionals in New Zealand sharing a common goal of up skilling New Zealand's unemployed people and graduates on various practical IT skills.

- We are Get Skills a New Zealand based IT training company.

- We are determined and committed for your success in your career path.

- We offer real time training to the candidates in the latest IT Trends like QA Testing Tools, Networking and certifications

- The courses being taught by the industry experts like IT managers and QA leads of Top MNC companies to share their real knowledge with you.

- Our objective is to Teach, Train and Place.

- Your Placement is our Commitment.

Contact US:

Yadwinder Sharma
C.E.O. & Founder – GetSkills Ltd New Zealand
Email: ceo@getskills.co.nz
Mobile: 00642102336373

Web: www.getskills.co.nz

www.ingramcontent.com/pod-product-compliance
Lightning Source LLC
Chambersburg PA
CBHW051912170526
45168CB00001B/357